AF595106

Contents

About the Editors vii

Rajesh Sunasee and Karina Ckless
Synthesis, Applications and Biological Impact of Nanocellulose
Reprinted from: *Nanomaterials* **2022**, *12*, 3188, doi:10.3390/nano12183188 1

Edmond Lam and Usha D. Hemraz
Preparation and Surface Functionalization of Carboxylated Cellulose Nanocrystals
Reprinted from: *Nanomaterials* **2021**, *11*, 1641, doi:10.3390/nano11071641 5

Abraham Samuel Finny, Oluwatosin Popoola and Silvana Andreescu
3D-Printable Nanocellulose-Based Functional Materials: Fundamentals and Applications
Reprinted from: *Nanomaterials* **2021**, *11*, 2358, doi:10.3390/nano11092358 37

Hamid M. Shaikh, Arfat Anis, Anesh Manjaly Poulose, Niyaz Ahamad Madhar and Saeed M. Al-Zahrani
Date-Palm-Derived Cellulose Nanocrystals as Reinforcing Agents for Poly(vinyl alcohol)/Guar-Gum-Based Phase-Separated Composite Films
Reprinted from: *Nanomaterials* **2022**, *12*, 1104, doi:10.3390/nano12071104 61

Alexandre Bernier, Tanner Tobias, Hoang Nguyen, Shreshth Kumar, Beza Tuga, Yusha Imtiaz, Christopher W. Smith, Rajesh Sunasee and Karina Ckless
Vascular and Blood Compatibility of Engineered Cationic Cellulose Nanocrystals in Cell-Based Assays
Reprinted from: *Nanomaterials* **2021**, *11*, 2072, doi:10.3390/nano11082072 75

Renato Mota, Ana Cristina Rodrigues, Ricardo Silva-Carvalho, Lígia Costa, Daniela Martins, Paula Sampaio, Fernando Dourado and Miguel Gama
Tracking Bacterial Nanocellulose in Animal Tissues by Fluorescence Microscopy
Reprinted from: *Nanomaterials* **2022**, *12*, 2605, doi:10.3390/nano12152605 87

Max Wacker, Jan Riedel, Heike Walles, Maximilian Scherner, George Awad, Sam Varghese, Sebastian Schürlein, Bernd Garke, Priya Veluswamy, Jens Wippermann and Jörn Hülsmann
Comparative Evaluation on Impacts of Fibronectin, Heparin–Chitosan, and Albumin Coating of Bacterial Nanocellulose Small-Diameter Vascular Grafts on Endothelialization In Vitro
Reprinted from: *Nanomaterials* **2021**, *11*, 1952, doi:10.3390/nano11081952 107

About the Editors

Rajesh Sunasee

is an Associate Professor of Organic Chemistry at the State University of New York at Plattsburgh. He received his Ph.D. in Organic Chemistry in 2009 from the University of Alberta, Canada, and was the recipient of an NSERC-IRDF industrial post-doctoral fellowship at OmegaChem Inc., Quebec. His research primarily focuses on the design, synthesis and characterization of functional advanced cellulose nanocrystal materials for potential biomedical applications such as vaccine nanoadjuvants, antimicrobial and anti-cancer agents. His research has been disseminated in 27 peer-reviewed publications, 6 book chapters and over 30 published conference abstracts at national and international scientific meetings. He is the recipient of the prestigious SUNY Chancellor Award for Excellence in Teaching and, recently, the Plattsburgh Alumni's Faculty Impact Award. He also serves as a Racial, Equity and Justice Fellow and an Equity Advocate member at SUNY Plattsburgh. He is currently the Councilor of the Northern New York ACS Local Section.

Karina Ckless

Karina Ckless is a Professor of Biochemistry at the State University of New York at Plattsburgh and is the recipient of the SUNY Chancellor's Award for Excellence in scholarship and creative activity. She received her PhD in Biochemistry from the UFRGS, Brazil. Her research involves the study of reactive oxygen species (ROS) and their impact on cell signaling, especially during inflammation and immune response. More recently, her work focus on the cytotoxicity and immunomodulatory activity of cellulose-based nanomaterials and their potential application as vaccine adjuvants. She has published over 40 peer-reviewed publications, 4 book chapters and several conference abstracts at national and international scientific meetings. Currently, she is a council member of the Society for Redox Biology and Medicine (SfRBM).

 MDPI

Editorial

Synthesis, Applications and Biological Impact of Nanocellulose

Rajesh Sunasee * and Karina Ckless *

Department of Chemistry and Biochemistry, State University of New York at Plattsburgh, Plattsburgh, NY 12901, USA
* Correspondence: rajesh.sunasee@plattsburgh.edu (R.S.); kckle001@plattsburgh.edu (K.C.)

Citation: Sunasee, R.; Ckless, K. Synthesis, Applications and Biological Impact of Nanocellulose. *Nanomaterials* **2022**, *12*, 3188. https://doi.org/10.3390/nano12183188

Received: 31 August 2022
Accepted: 12 September 2022
Published: 14 September 2022

Interest in cellulose-based nanomaterials has continued to increase dramatically in the past few years, especially with advances in the production routes of nanocellulose—such as cellulose nanocrystals (CNC), cellulose nanofibrils (CNF) and bacterial nanocellulose (BNC)—that tailor their performances [1–3]. In addition, given the presence of ample hydroxyl groups on the surface of nanocellulose, there have been significant developments in the chemical modifications of nanocellulose toward the design, fabrication and characterization of advanced functional nanocellulose-based materials for various applications [4–7]. Nanocellulose exhibits unique characteristics (v/s molecular cellulose) due to its nanoscale size, large surface area and other physicochemical properties. Research in the past decade has shown that nanocellulose is a promising bio-based material for several biomedical applications [6,8], given its biocompatibility and relatively low risk to human health. However, to advance the nano-biomedical applications of nanocellulose, it is crucial to develop a solid understanding of the biological impacts of nanocellulose, including toxicity, genotoxicity and potential immune responses elicited by these nanomaterials [9].

The present Special Issue in Nanomaterials aims to highlight recent advances in the synthesis of nanocellulose, surface modifications for the design of functional nanocellulose as well as its applications and potential biological impact. It features two review articles and four original research articles authored and reviewed by experts in the nanocellulose field. The compilation of the reviews and research articles for this book targets a broad readership of chemists, materials scientists, biochemists, nanotechnologists and others with an interest in nanocellulose research.

An extensive review by Lam and Hemraz discusses the preparation and surface functionalization of carboxylated CNC [10]. While sulfated CNC, derived from the sulfuric-acid-mediated hydrolysis of cellulose, has been the predominant form of this class of nanocellulose, carboxylated CNC has emerged as a similar material of interest for various studies and potential applications. The presence of the carboxyl groups on the nanocrystal's surface enables several chemical modification approaches to be explored for the design of functionalized carboxylated CNC. This review targets the recent progress in methods and feedstock materials for producing carboxylated CNC, their functional properties and discussions on the initial successes in their applications. The authors also discuss some of the inherent advantages that carboxylated CNC might possess in similar applications with sulfated CNC. The second review by Finny et al. focuses on 3D-printable nanocellulose-based functional materials [11]. The authors summarize the potential of nanocellulose as a promising material for designing functional nanostructures and devices via 3D printing. The different properties, preparation methods, printability and strategies to functionalize nanocellulose into 3D-printed constructs are thoroughly discussed. The recent development in 3D-printed nanocellulose-based composites for food, environmental, food packaging, energy, and electrochemical applications is also highlighted.

In an original research study, Shaikh et al. report the preparation of poly(vinyl alcohol)/guar-gum-based phase-separated film, which is then incorporated with date-palm-derived CNC as the reinforcing agent [12]. The films are synthesized via the solution-casting method and characterized by various spectroscopic and microscopic techniques.

A drug release study shows that the release of Moxifloxacin can be tailored by altering the CNC content within the phase-separated films. Overall, the developed films are found to be ideal as delivery carriers for Moxifloxacin. The biological impact of nanocellulose is portrayed in three other original research studies [13–15]. For instance, Bernier et al. investigated the impact of a library of cationic CNC in the human blood and endothelial cells using cell-based assays [13]. They observe that despite the cationic CNC not changing RBC morphology or causing aggregation, at 24 h exposure, mild hemolysis is detected, mainly with pristine CNC. They further study the effect of various concentrations of CNC on the cell viability of human umbilical vein endothelial cells (HUVECs) in a time-dependent manner. Results indicated that none of the cationic CNCs caused a dose–response decrease in the cell viability of HUVEC at 24 h or 48 h of exposure. Overall, the findings of this study, combined with the authors' previous study on the immunomodulatory properties of these cationic CNCs [16], support the potential development of engineered cationic CNCs as vaccine nanoadjuvants.

In another study, Mota et al. describe the use of the fluorescence microscopy technique to detect and visualize BNC and BCNC nanomaterials [14]. Both BNC and BCNC are prepared and characterized, and using adsorption studies, the interaction of a cellulose-binding module fused to a green fluorescent protein (GFP–CBM) with BNC and BCNC is investigated along with the uptake of BCNC by macrophages. An initial in vivo study shows that BNC or BCNC throughout the gastrointestinal tract is only observed in the intestinal lumen, indicating that cellulose particles are not absorbed. Wacker et al. study the effect of surface-coating BNC small-diameter vascular grafts with human albumin, fibronectin or heparin–chitosan upon endothelialization with human saphenous vein endothelial cells (VEC) or endothelial progenitor cells (EPC) in vitro [15]. The results indicate that the fibronectin coating significantly promotes the adhesion and growth of VEC and EPC, while albumin only promotes the adhesion of VECs. The heparin–chitosan coating only significantly improves the adhesion of EPC. Overall, both fibronectin and heparin–chitosan coatings could impact the endothelialization of BNC-SDVGs and, as such, could be promising approaches to help improve the longevity and reduce the thrombogenicity of BNC small-diameter vascular grafts.

Overall, this book compiles two excellent reviews with highlights of challenges and future directions as well as four articles related to research with nanocellulose from the Special Issue entitled "Synthesis, Applications and Biological Impact of Nanocellulose".

Funding: R.S. and K.C. acknowledged funding by the National Science Foundation under grant number 1703890.

Acknowledgments: The Guest Editors acknowledge all the authors for their contributions to this Special Issue, the reviewers for their dedicated time and expertise in reviewing the submitted papers and the editorial staff of Nanomaterials for their support throughout the timeline of this Special Issue.

Conflicts of Interest: The authors declare no conflict of interest.

References

1. Klemm, D.; Kramer, F.; Moritz, S.; Lindstroem, T.; Ankerfors, M.; Gray, D.; Dorris, A. Nanocelluloses: A New Family of Nature-Based Materials. *Angew. Chem. Int. Ed.* **2011**, *50*, 5438–5466. [CrossRef] [PubMed]
2. Vanderfleet, O.M.; Cranston, E.D. Production routes to tailor the performance of cellulose nanocrystals. *Nat. Rev. Mater.* **2021**, *6*, 124–144. [CrossRef]
3. Reid, M.S.; Villalobos, M.; Cranston, E.D. Benchmarking Cellulose Nanocrystals: From the Laboratory to Industrial Production. *Langmuir* **2017**, *33*, 1583–1598. [CrossRef] [PubMed]
4. Trache, D.; Tarchoun, A.F.; Derradji, M.; Hamidon, T.S.; Masruchin, N.; Brosse, N.; Hussin, M.H. Nanocellulose: From fundamentals to advanced applications. *Front. Chem.* **2020**, *8*, 392. [CrossRef] [PubMed]
5. Sunasee, R. Nanocellulose: Preparation, Functionalization and Applications. In *Comprehensive Glycoscience*, 2nd ed.; Barchi, J.J., Jr., Ed.; Elsevier: Amsterdam, The Netherlands, 2021; Volume 4, pp. 506–537.
6. Sunasee, R.; Hemraz, U.D.; Ckless, K. Cellulose nanocrystals: A versatile nanoplatform for emerging biomedical applications. *Expert Opin. Drug. Deliv.* **2016**, *13*, 1243–1256. [CrossRef] [PubMed]

7. Foster, E.J.; Moon, R.J.; Agarwal, U.P.; Bortner, M.J.; Bras, J.; Camarero-Espinosa, S.; Chan, K.J.; Clift, M.J.; Cranston, E.D.; Eichhorn, S.J.; et al. Current characterization methods for cellulose nanomaterials. *Chem. Soc. Rev.* **2018**, *47*, 2609–2679. [CrossRef] [PubMed]
8. Yuan, Q.; Bian, J.; Ma, M.G. Advances in Biomedical Application of Nanocellulose-Based Materials: A Review. *Curr. Med. Chem.* **2021**, *28*, 8275–8295. [CrossRef] [PubMed]
9. Čolić, M.; Tomić, S.; Bekić, M. Immunological aspects of nanocellulose. *Immunol. Lett.* **2020**, *222*, 80–89. [CrossRef] [PubMed]
10. Lam, E.; Hemraz, U.D. Preparation and Surface Functionalization of Carboxylated Cellulose Nanocrystals. *Nanomaterials* **2021**, *11*, 164111. [CrossRef] [PubMed]
11. Finny, A.S.; Popoola, O.; Andreescu, S. 3D-Printable Nanocellulose-Based Functional Materials: Fundamentals and Applications. *Nanomaterials* **2021**, *11*, 2358. [CrossRef] [PubMed]
12. Shaikh, H.M.; Anis, A.; Poulose, A.M.; Madhar, N.A.; Al-Zahrani, S.M. Date-Palm-Derived Cellulose Nanocrystals as Reinforcing Agents for Poly(vinyl alcohol)/Guar-Gum-Based Phase-Separated Composite Films. *Nanomaterials* **2022**, *12*, 1104. [CrossRef] [PubMed]
13. Bernier, A.; Tobias, T.; Nguyen, H.; Kumar, S.; Tuga, B.; Imtiaz, Y.; Smith, C.W.; Sunasee, R.; Ckless, K. Vascular and Blood Compatibility of Engineered Cationic Cellulose Nanocrystals in Cell-Based Assays. *Nanomaterials* **2021**, *11*, 2072. [CrossRef] [PubMed]
14. Mota, R.; Rodrigues, A.C.; Silva-Carvalho, R.; Costa, L.; Martins, D.; Sampaio, P.; Dourado, F.; Gama, M. Tracking Bacterial Nanocellulose in Animal Tissues by Fluorescence Microscopy. *Nanomaterials* **2022**, *12*, 2605. [CrossRef] [PubMed]
15. Wacker, M.; Riedel, J.; Walles, H.; Scherner, M.; Awad, G.; Varghese, S.; Schürlein, S.; Garke, B.; Veluswamy, P.; Wippermann, J.; et al. Comparative Evaluation on Impacts of Fibronectin, Heparin–Chitosan, and Albumin Coating of Bacterial Nanocellulose Small-Diameter Vascular Grafts on Endothelialization In Vitro. *Nanomaterials* **2021**, *11*, 1952. [CrossRef] [PubMed]
16. Imtiaz, Y.; Tuga, B.; Smith, C.W.; Rabideau, A.; Nguyen, L.; Liu, Y.; Hrapovic, S.; Ckless, K.; Sunasee, R. Synthesis and Cytotoxicity Studies of Wood-Based Cationic Cellulose Nanocrystals as Potential Immunomodulators. *Nanomaterials* **2020**, *10*, 1603. [CrossRef] [PubMed]

MDPI

Review

Preparation and Surface Functionalization of Carboxylated Cellulose Nanocrystals

Edmond Lam and Usha D. Hemraz *

Aquatic and Crop Resource Development Research Centre, National Research Council of Canada, Montreal, QC H4P 2R2, Canada; Edmond.Lam@nrc-cnrc.gc.ca
* Correspondence: Usha.Hemraz@nrc-cnrc.gc.ca

Abstract: In recent years, cellulose nanocrystals (CNCs) have emerged as a leading biomass-based nanomaterial owing to their unique functional properties and sustainable resourcing. Sulfated cellulose nanocrystals (sCNCs), produced by sulfuric acid-assisted hydrolysis of cellulose, is currently the predominant form of this class of nanomaterial; its utilization leads the way in terms of CNC commercialization activities and industrial applications. The functional properties, including high crystallinity, colloidal stability, and uniform nanoscale dimensions, can also be attained through carboxylated cellulose nanocrystals (cCNCs). Herein, we review recent progress in methods and feedstock materials for producing cCNCs, describe their functional properties, and discuss the initial successes in their applications. Comparisons are made to sCNCs to highlight some of the inherent advantages that cCNCs may possess in similar applications.

Keywords: carboxylated cellulose nanocrystals; functionalization; surface treatment; nanomaterial; biomaterial

Citation: Lam, E.; Hemraz, U.D. Preparation and Surface Functionalization of Carboxylated Cellulose Nanocrystals. *Nanomaterials* **2021**, *11*, 1641. https://doi.org/10.3390/nano11071641

Academic Editor: Hirotaka Koga

Received: 29 May 2021
Accepted: 17 June 2021
Published: 22 June 2021

1. Introduction

Interest in biomass-based nanomaterials has grown in the last decade owing to their unique functional properties and sustainable resourcing. Currently, cellulose nanocrystals (CNCs) are among the leading biomass-based nanomaterials in terms of publications, applications developed, and technology readiness level (TRL). First produced by Nickerson and Habrle in 1947 [1], CNCs are crystalline, rod-shaped particles ranging in size from 5 to 20 nm in width and hundreds of nm in length, depending on the biomass source and production method. In order to obtain CNCs, native semi-crystalline cellulose is broken down into its elementary crystalline domains with the concurrent removal of amorphous cellulose segments. With its high aspect ratio [2], colloidal stability in aqueous media [3], liquid crystalline properties [4], and biocompatibility [5], CNCs have been used in a diverse range of applications, including but not limited to polymer composites [6], electronics [7], biomedical [8], and photonic films [9]. A number of excellent review papers have been published over the years that document the technological advances related to CNC's research [10–16].

The most commonly used method for CNC's synthesis is via sulfuric acid-mediated hydrolysis, in which removal of the amorphous cellulose segments is facilitated by the hydrolysis of glycosidic bonds and concomitant esterification of surface hydroxyl groups to form sulfate half-ester groups [17]. Mukherjee and Woods found that acid concentration (64 wt. % sulfuric acid) was an important factor in producing sulfated cellulose nanocrystals (sCNCs) [18]. Today, these sCNCs are produced by a number of companies, including Alberta-Pacific Forest Industries, GranBio, and the industry-leading Celluforce, who utilize forestry pulp feedstocks [15]. At the moment, the leading industrial producers of cCNCs are located in Canada: Anomera (30 kg/day production, target 1 ton/day in 2021; hydrogen peroxide-assisted) and Blue Goose Biorefineries (10 kg/day production; transition-metal-catalyzed) [15]. These daily production rates are only at the pilot scale and lag behind the industry leader, Celluforce, which has a production capacity of 1 ton per day for sCNCs. Although sulfuric acid hydrolysis is the predominant method for producing CNCs, there are many other methods reported in the literature that allow for the preparation and isolation of CNCs at various levels of TRL. These methods include enzymes [19], oxidizers [20–22], mechanical treatments [23], or a combination of these means [24,25]. Efforts in developing alternative CNCs' production methods to sulfuric acid hydrolysis have been driven by the various challenges associated with using sCNCs in downstream applications, including low thermal stability [26] and nanomaterial aggregation [27].

Researchers have also turned to other methods to improve the economics and sustainability of the CNC production process. The sulfuric acid hydrolysis method is a harsh reaction due to the corrosiveness of the acid employed and favors the use of pure cellulosic feedstocks (forestry pulp materials, for example) to produce sCNCs. When biomass waste streams (such as raw lignocellulosic biomass containing cellulose, lignin, pectin, and other biopolymers) are used in the same process, it often leads to a low-yielding, impure CNC product that requires additional purification steps [20]. Alternative methods and different cellulosic feedstock materials have led to the production of CNCs with varying yields and different functional properties (morphology, surface chemistry, and surface charge density) compared to sCNCs. For example, other mineral acids such as HCl [28] and H_3PO_4 [29] have been used to produce CNCs, but these products have low surface charge and, as a result, do not disperse as well as the sCNCs. Carboxylated cellulose nanocrystals (cCNCs) have similar colloidal stability, uniform nanoscale lengths, and high crystallinity compared to sCNCs. Similar to sulfate half-ester groups of sCNCs, cCNCs possess carboxyl groups that promote the electrostatic repulsion between neighboring CNCs that prevents aggregation. One particularly enticing feature of cCNCs is the ability for these carboxyl groups to undergo further reactivity for surface modification to tailor the properties of the nanomaterial for downstream applications.

This review provides a summary of the different methods used to produce cCNCs from a diverse range of feedstock materials. With the availability of the carboxyl groups on the surface of the nanocrystal, a wide range of chemical modification approaches were explored to further alter the functional properties of the nanomaterial. As methods to produce cCNCs have become more mature, more applications utilizing cCNCs are emerging, and examples in the recent literature are also presented. Comparisons of cCNCs to sCNCs highlight the advantages the carboxyl group may bring to these end applications. Finally, a future outlook is provided to address future challenges and directions. This review is focused on the direct synthesis of cCNCs from biomass feedstocks and excludes reactions performed on nanocelluloses and other types of cellulose derivatives.

2. Production of cCNCs

It is possible to obtain cCNCs through four general classes of reactions (Scheme 1), and although they all feature the carboxyl functional group on the surface of CNCs, the location of the carboxyl groups differ slightly:

1. Carboxyl group on the C6 position of the cellulose chain, generally obtained through oxidation of the primary hydroxyl groups at the C6 position.
2. Carboxylic acid group tethered to a moiety covalently attached to the hydroxyl group on the C6 position in addition to carboxylic acid group on the C6 position when considering two anhydroglucose units (AGU), typically obtained through the esterification of cellulose hydroxyl groups with carboxylic acids.
3. 2,3-Dicarboxylic acids from glucose ring opening.
4. Carboxylic acid groups found exclusively at the reducing ends of CNCs, which can be obtained due to the chemistry of the highly reactive aldehyde functional group.

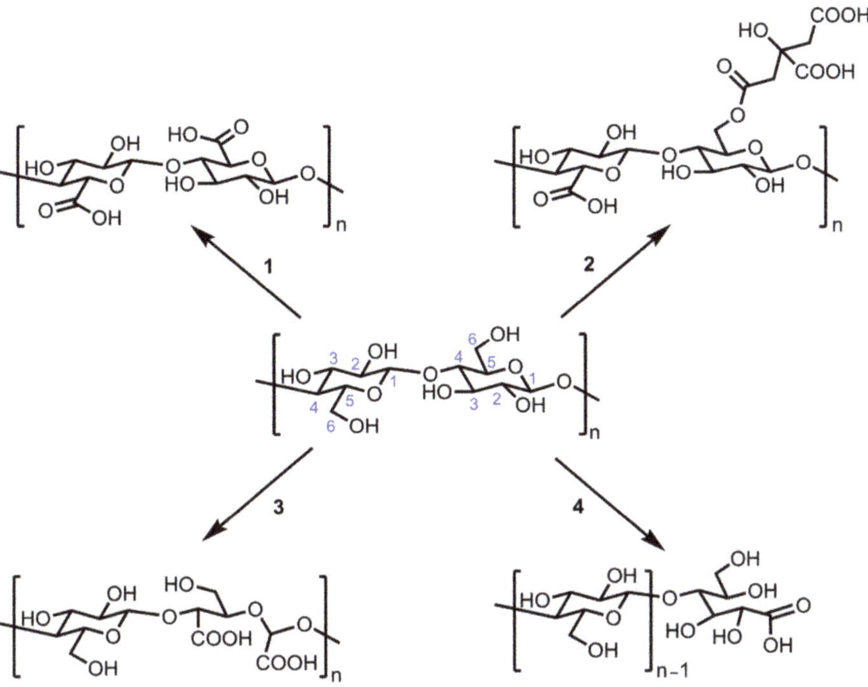

Scheme 1. Four main types of cCNCs, produced through reactions on various cellulosic sources and utilized for a wide range of applications.

2.1. Class 1: Carboxyl Group on the C6 Position of the Cellulose Chain

Among the four classes of cCNCs depicted in Scheme 1, most of the widely used methods produce cCNCs containing carboxyl groups at the C-6 position. It is possible to generate this type of cCNCs either through various oxidative hydrolysis processes or through a two-step methodology of acidic hydrolysis, followed by oxidation of cellulosic biomass. The most commonly used oxidizing agents for the generation of cCNCs with carboxyl groups at the C-6 position are shown in Scheme 2.

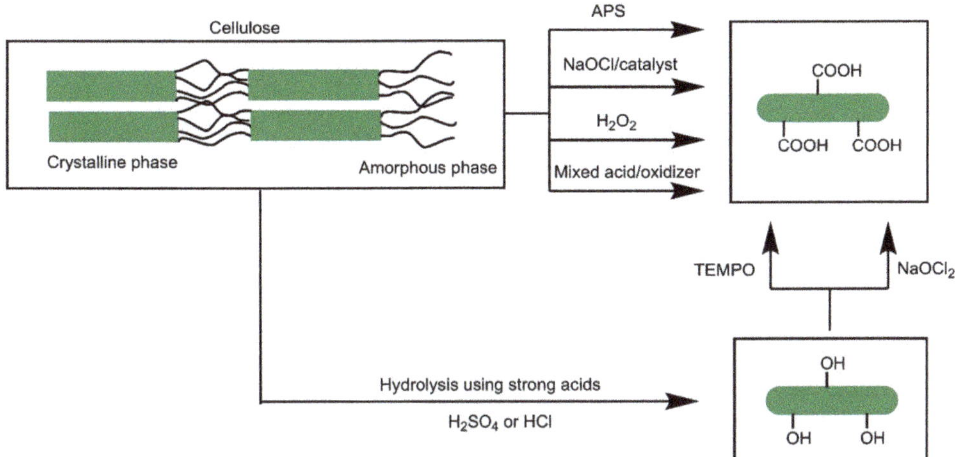

Scheme 2. General methods for obtaining cCNCs from cellulosic sources.

2.1.1. TEMPO

The 2,2,6,6-tetramethyl-1-piperidinyloxy (TEMPO)-mediated oxidation reaction is the most popular way to incorporate carboxylic acid groups on the surface of the nanocrystals (Scheme 3) through the oxidation of the primary hydroxyl groups of CNCs. In most cases, this method has been adopted as a post-production strategy on CNCs produced from the acid hydrolysis of cellulosic sources. The TEMPO oxidation was first introduced by De Nooy et al., whereby primary hydroxyl groups on polysaccharides were selectively oxidized in the presence of the secondary hydroxyl groups [30]. Its applicability on CNCs was first reported by Vignon and co-workers on nanocrystals produced from the HCl hydrolysis of cotton linter, subjected to a mixture of sodium hypochlorite, sodium bromide, and TEMPO for the selective oxidation of the primary surface hydroxyl groups to produce cCNCs. The carboxyl content or the degree of oxidation (DO) can be measured by conductometric titrations, FT-IR, and methylene blue adsorption, with conductometry being the most reliable method. In this case, a DO of 0.15 was obtained by conductimetry [22]. Due to the presence of negative surface charges, the cCNCs formed are well-dispersed in water and produce non-flocculating birefringent suspensions. It was noted, however, that excessive TEMPO oxidation led to a decrease in the crystal size. As such, the authors next optimized the oxidation process by varying the ratio of NaOCl to AGU such that only the accessible primary hydroxyl groups on the surface of the nanocrystals would be oxidized without affecting the core. The remaining hydroxyl groups are not affected by the oxidation process–this includes the inaccessible half of the primary hydroxyl groups on the surface, secondary hydroxyl groups on the surface, and all hydroxyl groups of the glucose units in the core crystal (see Scheme 4). This is consistent with other cCNCs of this class, where the carboxyl group is located exclusively on the C6 position of the cellulose chain. The TEMPO-oxidized product is often used as a precursor for further chemical modification via the reactive carboxylate group. While this review is not an exhaustive account of all TEMPO-mediated oxidation reactions and their subsequent transformations, some examples of further chemical functionalization and applications are discussed in later sections.

Scheme 3. TEMPO oxidation of CNCs [22].

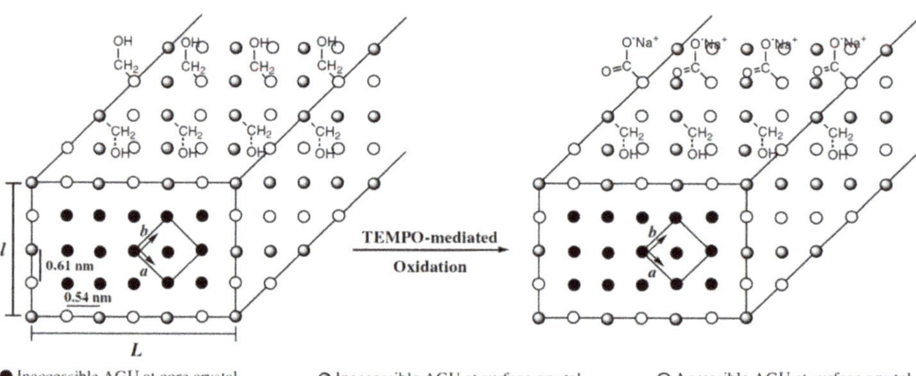

Scheme 4. TEMPO oxidation of accessible hydroxyl groups at the surface of the CNCs (Reproduced with permission from [31]. Copyright Springer Science Business Media, Inc. 2006).

As opposed to using TEMPO oxidation to introduce carboxyl groups after CNCs' production, Isogai and co-workers initially treated softwood bleached kraft pulp (SBKP) and microcrystalline cellulose (MCC) with a TEMPO mixture system [24]. After surface oxidation of the primary hydroxyl groups, the oxidized cellulose materials were exposed to cavitation-induced forces by sonication in water for 10–120 min to produce cCNCs with an average length of 200 nm and 3.5–3.6 nm in width. cCNCs prepared from SBKP exhibited higher mass recovery ratios and carboxylate content compared to those produced from MCC. Furthermore, the cCNCs prepared by TEMPO-sonomechanical treatments had a more uniform size distribution and exhibited a higher negative surface charge compared to CNCs prepared by acid hydrolysis, and did not require further purification through dialysis.

Faria and co-workers produced cCNCs from elephant grass by milling and hydrothermal processing with H_2SO_4 to remove hemicellulose and other extractives and NaOH to remove lignin [32]. This pre-treated feedstock was oxidized by TEMPO and mechanically disintegrated by sonication to yield cCNCs. They demonstrated that cCNCs are potential anti-biofouling agents that utilize contact-mediated membrane stress as the mechanism governing the toxicity of CNCs towards bacteria cells. In follow-up work, the same elephant grass cCNCs were crosslinked with polyamide membranes to produce antimicrobial thin-film composites [33].

2.1.2. Ammonium Persulfate (APS)

The TEMPO process is a powerful reaction to convert surface hydroxyl groups of CNCs to carboxyl groups, and yet it has several limitations. To produce cCNCs, either acid hydrolyzed CNCs are subjected to a post-production surface modification with TEMPO [22], or TEMPO-oxidized cellulosic materials are further treated with acid hydrolysis [31,34] or sonomechanical treatments [24]. The TEMPO process cannot directly produce cCNCs from longer-chain cellulosic materials as it is unable to break down the amorphous domains of cellulose [35]. To this end, other non-acid methods have been sought for the production of CNCs with similar potency to mineral acids. Despite the low cost and high abundance of mineral acids like sulfuric acid, these production methods require expensive corrosion-resistant equipment, multiple treatments, and isolation steps such as dialysis. In 2011, Luong and co-workers developed a process to produce cCNCs using APS, a low-cost, low-toxicity oxidant, from a variety of native plant materials and bacterial cellulose [20]. The cellulosic materials were heated at 60 °C in 1 M APS for 16 h with vigorous stirring. Under these conditions, the persulfate anions undergo two reactions in solution:

$$S_2O_8{}^{2-} + \text{heat} \rightarrow SO_4{}^{\cdot -} \tag{1}$$

$$S_2O_8{}^{2-} + 2H_2O \rightarrow 2HSO_4{}^{-} + H_2O_2 \tag{2}$$

The free radicals and hydrogen peroxide are capable of hydrolyzing the amorphous cellulose, decolorizing the product via the oxidative breakdown of the aromatic components of the plant material, and oxidizing the C6 alcohol groups on the surface of the nanocrystal. The yields of the cCNCs vary greatly as they depend on the starting biomass materials: native plant materials (such as flax, hemp, and triticale) yield lower amounts of cCNCs compared to MCC or paper filters as the plant materials possess additional components such as lignin, hemicellulose, and wax, which lower the weight percentage of cellulose in the starting material. The DO of the resulting cCNCs ranged from 0.11 to 0.19. From TGA analysis, cCNCs produced were found to be more thermally stable than sCNCs. To better understand the nature of the APS reagent breakdown, Lam et al. conducted Raman spectroscopy studies on the reaction mixtures of APS and MCC and found that nearly 60% of the sulfate ions in solution were attributed to the H_2SO_4 [36]. By the addition of NH_4OH to the solution, it was possible to not only initiate the recovery of sulfate anions as ammonium sulfate but the resulting neutralized cCNCs with $COO^-NH_4{}^+$

groups exhibited better dispersion and thermal characteristics over cCNCs with COOH and COO^-Na^+ groups.

A limitation to the APS method by Luong and co-workers is the large amount of APS required for biomass conversion (22.8 g APS per g of biomass) and the long reaction time (minimum of 16 h) [20]. Liu and co-workers investigated the use of *N*,*N*,*N*′,*N*′-tetramethylethylenediamine (TMEDA) and ultrasonic-assisted disintegration to address the economic issues of the APS oxidation [37]. Under the conventional APS method, a single type of free radical is produced ($SO_4\cdot^-$), which Liu et al. believe results in the relatively weak hydrolysis ability of APS, necessitating a higher loading of the oxidant and longer reaction times to produce cCNCs. By using TMEDA as a redox initiator, two additional free radicals ($SO_4\cdot^-$ and $HO\cdot$) were also produced that can react with the biomass at higher redox potentials [38]. An ultrasonic step was further added to the process to promote disintegration of the cellulosic chain by increasing the surface area for increased reaction sites. Under these conditions, cotton pulp was successfully converted to cCNCs in 6 h at 75 °C for a yield of 62.5%, with lower consumption of APS (8.5 g APS per g of pulp).

Amoroso et al. examined the use of a microwave-assisted process to overcome the inefficiencies of heating currently used in the conventional APS process [39]. They found that controlling the heating ramp time and holding the heating time were important parameters to allow for sufficient time for the APS to convert to the active radical oxidant reacting with the biomass. Limiting microwave power output also reduced the potential of carbonizing the cellulosic material, which was detrimental to the cCNC's yield. Cotton residues were converted to cCNCs in 90 min under microwave-assisted heating, which demonstrates the process as an effective way to reduce the reaction time of the APS process. However, the limitation to the microwave-assisted reaction comes from scalability: APS conventional heating can be performed at 1000 L, while the current microwave-assisted method has only been demonstrated at 50 mL volume.

Researchers are often interested in comparing hydrolysis methods for producing CNCs from a variety of waste streams, whether they are post-consumer products, agricultural residues, or unlikely non-plant sources of cellulosic materials. For example, recycled paper waste is often viewed as a source of secondary cellulosic fibers for valorization. Jiang et al. evaluated the conversion of these materials into sCNCs and cCNCs (by APS oxidation) [40]. cCNCs were produced in lower yield (22.42%) compared to sCNCs (41.22%), with a crystallinity index (CrI) of 77.56% and 72.45%, respectively. It was noted that sulfuric acid residues on the surface of the sCNCs could promote the carbonization of cellulose and increase the production of carbon residues. The carbon residue of sCNCs was 26.73% compared to only 14.17% for cCNCs. Zhang and co-workers investigated different methods to produce CNCs from lemon (*Citrus limon*) seeds, including sulfuric acid treatment for sCNCs, and both APS and TEMPO oxidation for cCNCs [41]. In all three methods, the CNCs retained their cellulose Iβ structure. TEMPO-oxidized cCNCs produced the highest yield of cCNCs with larger nanorod dimensions (~360 × 34 nm) and lower CrI compared to the other two methods. APS-treated cCNCs had the highest CrI and sCNCs the smallest nanocrystal dimensions. The lemon-seed-derived CNCs were tested in Pickering emulsions, in which the sCNCs and APS-treated cCNCs exhibited the best stabilizing effects compared to the TEMPO-oxidized cCNCs.

The use of APS to convert bamboo borer powder (produced when *lyctus bruneus*, a beetle, attacks the woody bamboo materials to yield a flour-like powder) to cCNCs as a means to valorize bamboo raw materials was investigated [42]. The resulting cCNCs were more spherical in shape (20–50 nm) and had a CrI of 62.75%. The spherical geometry of these particles could be attributed to the nanoparticles and was possibly created by a self-assembly process of CNCs and their fragments from the less-crystalline cellulose sources [43]. Lu and Hsieh also produced spherical sCNCs from chardonnay grape skins in which the starting cellulose's CrI increased from 54.9% to 64.3%. The authors believe that the lower crystallinity of the grape skin cellulose produces fewer rod-like crystals compared to cotton and wood, which have larger intact crystalline regions. Furthermore, the drying process could promote the aggregation of smaller, more abundant, non-rod nano-fragments around less abundant, larger nano-rods by hydrogen bonding, leading to a core–shell spherical cellulose nanostructure.

Tunicates are marine invertebrate sea animals that are the only known animal source of cellulose. Cellulose from the mantle of *Halocynthia roretzi* was extracted through several alkali treatments and bleaching processes. The tunicate cellulose was then treated with APS and ultrasonic post-processing to produce cCNCs [44]. The formation of tunicate cCNC lyotropic chiral nematic liquid crystals was observed for the first time, which displayed birefringence and a fingerprint texture. The critical concentration of phase separation for tunicate cCNC suspension was around 3.5 wt%. Solid films were subsequently prepared from evaporation-induced self-assembly, which showed preservation of the chiral nematic structure of tunicate cCNCs. In previous work using TEMPO-oxidized tunicate CNCs, only non-uniform birefringence was observed, which may be attributed to the high polydispersity of the length (from 0.1 to 10 μm) of the tunicate CNCs and their high viscosity suspensions [31].

Pan and co-workers converted bleached kraft pulp (cellulose I feedstock) into cellulose II using a mildly acidic lithium bromide trihydrate (MALBTH) system to induce the partial hydrolysis and polymorph transition [45]. APS was then used to convert the cellulose II into cCNCs II, which were found to be smaller in size compared to those of cellulose I cCNCs that were not subject to the polymorph transition procedure. Other researchers have focused on different feedstocks to produced cCNCs by the APS method, including recycled medium-density fiberboard [46], denim waste [47], and balsa and kapok fibers [48].

2.1.3. Hydrogen Peroxide

One of the simplest and greenest oxidizing agents available is hydrogen peroxide. Anomera, a company based in Quebec, Canada, has developed a process to produce cCNCs in one step from biomass, dissolving pulp and wood waste using 30% hydrogen peroxide at 115 °C for about 8 h using a method developed by Andrews and Morse [21]. An advantage to this process is the full consumption of hydrogen peroxide. In the same patent, it was also possible to produce cCNCs from spruce fibers at room temperature by irradiating a 30% hydrogen peroxide solution with UV light for 12 h. In addition, they were able to convert the negatively charged cCNCs to a positively charged nanomaterial by mixing the cCNCs with polydiallyldimethylammonium chloride (PDDA), a cationic polymer that is used as a wet-end additive in papermaking. The company's current cCNC product, Dextracel, has a size of 150–250 nm in length, a width of 5–10 nm, crystallinity of >85%, and a carboxyl content of 0.12–0.20 mmol/g.

2.1.4. Sodium Hypochlorite (NaOCl)/Sodium Chlorite ($NaOCl_2$)

Blue Goose Biorefineries is a company in Saskatchewan, Canada, that produces cCNCs from a variety of biomasses using a transition metal-catalyzed oxidation process [49]. The process is characterized by three steps. In the first step, a redox reaction using sodium hypochlorite and a catalyst (either iron or cupric sulfate) is employed to break down the starting biomass. The isolated biomass filter cake is then subjected to a sodium hydroxide treatment as part of the second step. In the final step, the alkaline-treated biomass is exposed to a second redox reaction in conditions similar to the first step, which finally results in the production of cCNCs. The yields of the reaction are dependent on the biomass source used. For example, the highest yields were reported for A96 high purity cellulose at 38.8% yield, but only 9.7% yield was reported for Yreka, a medium-density fiberboard with high lignin content. Despite similar reaction times required for all biomass materials tested, the amount of sodium hypochlorite required to produce cCNCs using this method increases with the lignin content in the starting biomass. The company's current cCNC product, BGB Ultra, has a size of 100–150 nm in length, a width of 9–14 nm, crystallinity of 80%, and a carboxyl content of 0.15 mmol/g.

The pilot-scale nanocellulose production at USDA Forest Product Laboratory utilizes a combination of sulfuric acid, sodium chlorite, and 4% hypochlorite solution to produce 25 kg of cCNCs per batch, with nanocrystals of 5–20 nm in width and 150–200 nm in length [50]. Using a 50 kg machine-dried pre-hydrolysis kraft rayon-grade dissolving pulp into a 400 L reactor, the cellulose strips are subjected to sulfuric acid hydrolysis at 45° C for 90 min (300 L, 64 wt. %) under a nitrogen atmosphere. Upon quenching the reaction, the acidic suspension is treated with sodium chlorite, sodium hydroxide, and 4% hypochlorite solution for neutralization and bleaching of the nanocrystals, which had been discolored by the sugar degradation during the acid hydrolysis process. Carboxyl groups are likely introduced when the suspension is treated with sodium chlorite and hypochlorite solutions.

2.1.5. Mixed Acid Solutions with Oxidizers

Some researchers have turned to using mixtures of acids and oxidizers to produce cCNCs. This strategy is intended to overcome some of the drawbacks that each reagent may contribute to the synthesis of the CNC product. For example, strong acids like sulfuric acid are capable of hydrolyzing cellulose to form CNCs, but the introduction of the sulfate ester groups may be undesirable as it may limit the utility of the CNCs by decreasing their thermal stability. Weak acids and oxidizers can be used to tune the surface chemistry of the nanocrystal but are unable to generate sufficiently high proton concentration to induce cellulose hydrolysis.

A one-pot procedure for cCNCs from cotton pulp was developed using 1% sulfuric acid, with potassium permanganate and oxalic acid, as the oxidizing and reducing agents, respectively [51]. cCNCs prepared from this method were 150–300 nm in length and 10–22 nm in width with a carboxyl content of 1.58 mmol/g. The authors believe that the oxalic acid can complex Mn^{3+} to form $[Mn(C_2O_4{}^{2-})]^+$ and prevent the Mn^{3+} from being reduced to Mn^{2+}, leading to the prolonged strong oxidizing capacity of the reaction system. Compared to sCNCs, their cCNCs' solutions of >6 wt. % displayed chiral nematic liquid crystalline phases.

A one-step hydrolysis process to convert MCC to cCNCs was developed using sulfuric acid and nitric acid in 0.5 h, whereby the resulting cCNCs had physicochemical characteristics consistent with those obtained from APS and TEMPO oxidation [52]. From 50 to 90 °C, the length and width of the nanocrystals decreased in size and yield while exhibiting increasing DO (up to a maximum of 0.11 at 80 °C). High crystallinity for these cCNCs (average CrI about 90%) was reported, though the starting MCC CrI was already at 85.3%.

2.2. *Class 2: Carboxylic Acid Group Tethered to a Moiety at C6 Position*

While sulfuric acid is the most common mineral acid employed for the production of CNCs, its use comes with economic and sustainability challenges. For example, in producing 1 kg of sCNCs, nearly 9 kg of sulfuric acid is consumed, leading to the generation of 13 kg of Na_2SO_4 from acid neutralization, with only moderate yields reported (30–50%) [53]. Organic acids, specifically carboxylic acids, are gaining traction as alternative hydrolysis reagents to mineral acids to overcome these economic, technical, and environmental challenges. However, a key criterion to produce CNCs is to use organic acids with sufficient acid strength (pKa = 1–3). For example, formic acid (pKa = 3.77) was used to hydrolyze birch pulp, resulting in mainly micron-length cellulosic fibers [54]. Only after increasing the acidity of the reaction mixture with 2% HCl were CNCs of <1000 nm produced. In addition, the use of formic acid did not produce cCNCs; a subsequent TEMPO oxidation step was required to convert micron-length cellulosic fibers into cCNCs.

Chen et al. demonstrated the use of different dicarboxylic acids (oxalic and maleic acid) on a bleached eucalyptus kraft pulp to produce CNCs and cellulose nanofibrils [53]. The carboxylation of the cellulose is not a direct oxidation of the C6 alcohol of cellulose (as in the case of TEMPO or APS oxidation), but a Fischer esterification, in which the organic acid is added onto the cellulose. The use of organic acids of relatively lower acid strength (pKa = 1.25 for oxalic acid; pKa = 1.9 for maleic acid) led to the production cCNCs of longer length and greater thermal stability (albeit at lower yields) compared to those produced from sulfuric acid, which has a higher acid strength (pKa = −3.0). In the optimal case of using 70% oxalic acid at 100 °C for 60 min, only a 25% yield for cCNCs was reported. The majority of the cellulosic residue left in the reaction mixture could be processed by mechanical fibrillation to produce CNF. Due to the low solubility of the organic acids in water, the acid was easily recovered by crystallization; nearly 95% of the oxalic acid used in the reaction was recovered from the hydrolysate, reflecting a greener process.

An extension into weak tricarboxylic acids was reported by using citric acid and ultrasonication methods to produce cCNCs from sugarcane bagasse pulp [55]. The addition of the post-ultrasonication method increased the yield of the cCNCs by 21.6% in comparison to the non-ultrasonication protocol. cCNCs were recovered by dialysis and centrifugation steps, and the remaining oxidized cellulosic residues were subjected to homogenization to produce cCNF (63.4%). Citric acid was subsequently recovered from the reaction liquor by rotary evaporation. A common problem cited with the use of citric acid is the low yield of cCNCs obtained due to the weak acidity of citric acid (pKa = 3.13), which is corroborated by the much higher isolated yield of cCNF compared to cCNCs. Higher yields (87.8%) of cCNCs were reported using 90% citric acid and 10% HCl to convert MCC to cCNCs, but this process required the addition of the strong mineral acid HCl to promote hydrolysis [56]. These citric-acid-oxidized cCNCs were then used in the preparation of oil/water emulsions for food applications. Emulsions made with higher concentrations of cCNCs (5%) resulted in smaller droplet sizes (1.01 μm) compared to those produced at lower concentrations of cCNCs (0.1%) at an average droplet size of 17.8 μm. The smaller droplets were more effective in stabilizing the soybean oil over 28 days. A similar approach (Scheme 5) has been utilized to produce dicarboxylated CNCs using the oxidative hydrolysis of MCC by APS, followed by esterification with citric acid in the presence of ultrasonication [57].

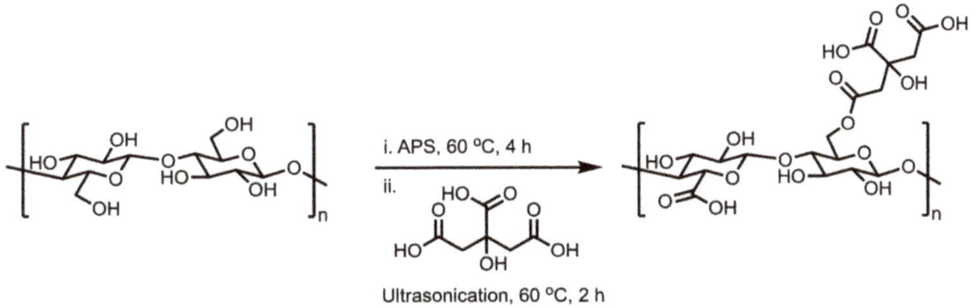

i. APS, 60 °C, 4 h

ii. Ultrasonication, 60 °C, 2 h

Scheme 5. Reaction of microcrystalline cellulose with APS and citric acid to produce CNCs with dicarboxylic acid moieties [57].

An underutilized plant in African nations is the cellulose-rich Juncus plant (*Juncus effuses*). The Juncus plant has a grass-like structure consisting of hollow cylindrical rods about 1 m in height, and 4–8 mm in diameter and are typically harvested as materials for woven textiles. Kassab et al. investigated the production of CNCs from the Juncus plant [58]. Purified cellulose microfibers (CMF) were first obtained from the plant stems, and subsequent treatment with citric acid/HCl hydrolysis produced cCNCs with a length of 352 ± 79 nm, a diameter of 6.1 ± 2.8 nm, and a CrI of 83% with cellulose I structure. The authors speculated that the Juncus plant cCNCs could be used as a nano-reinforcing agent for polymer composites due to the high degradation temperature (231 °C). In another report, the authors evaluated the use of post-harvest tomato plant residue (TPR) as a sustainable source for the extraction of cellulose derivatives, namely, CMF and CNC [59]. After obtaining CMF with an average diameter of 20 μm from the TPR, the CMF was then treated with a mixture of citric acid and HCl to produce the cCNCs. The resulting TPR cCNCs also possessed a cellulose I structure with a length of 514 ± 131 nm, a diameter of 4.7 ± 1.4 nm, and a CrI of 78%. A higher degradation temperature of 243 °C was reported for the TPR cCNCs compared to sCNCs and phosphorylated CNCs produced in the study (219 and 235 °C, respectively), a trend that was also observed for the Juncus plant CNCs.

Liu et al. were able to increase the cCNC's yield to 80.3% by using catalytic amounts of $FeCl_3$ to enhance citric acid hydrolysis efficiency [60]. The authors believe that $FeCl_3$ can promote the reaction in several ways: (1) as a Lewis acid to increase the acidity of the citric acid solution by polarizing the water molecules around the central metal ions, (2) the Fe^{3+} can coordinate with glucose to produce an intermediate complex with weakened C-O-C bonds that are more susceptible to bond breakage, and (3) disrupt inter- and intramolecular hydrogen bonds within cellulose to enhance the hydrolysis reaction.

2.3. Class 3: 2,3 Dicarboxylic Acids from Glucose Ring Opening

Carboxylation can be achieved by using sodium periodate, which can cleave the C2–C3 bonds of β-D-glucose monomer units of cellulose, and selectively oxidize C2 and C3 vicinal hydroxyl groups to form 2,3-dialdehyde units along the cellulose. These aldehyde groups can undergo further oxidation to form 2,3-dicarboxyl groups by sodium chlorite in an aqueous acidic medium (Scheme 6). This two step-process was used to convert softwood pulp into cCNCs [61]. Initial separation of the two fractions yielded mainly oxidized microfibrils. Upon alcohol addition, a second fraction of cCNCs was obtained with dicarboxylated functionalities on the surface of the nanomaterial and a carboxyl content ranging from 3.60 mmol/g to 6.60 mmol/g. Based on dynamical light scattering (DLS) measurements, these dicarboxylated cCNCs had significantly larger hydrodynamic radii, which the authors attribute to the presence of dicarboxylate chains protruding off the main CNC nanorods, contributing to the increased carboxyl content values obtained. Upon further hydrolysis of these dicarboxylated cCNCs, the carboxyl content of the materials decreased to 1.4 mmol/g as these dicarboxylate chains are more readily hydrolyzable compared to the crystalline CNC nanorods.

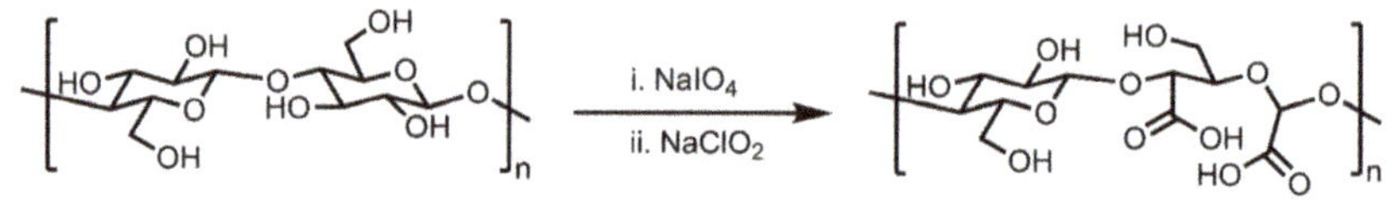

Scheme 6. Formation of cCNCs via oxidative cleavage of C2–C3 bonds using sodium periodate oxidation, followed by oxidation of the aldehyde groups into carboxyl groups using sodium chlorite [61].

Sugarcane bagasse was used to prepare cCNCs using two methods: a single-step sodium periodate oxidation method and the two-step sulfuric acid/TEMPO oxidation method [62]. One aspect of the investigation was to compare the physicochemical properties of the cCNCs of the different oxidation methods. When the sodium periodate reaction was performed at room temperature, rod-like cCNCs (104 × 6 nm) were obtained. However, at 60 °C, spherical nanoparticles (approx. 24 nm in diameter) were formed instead, which the authors attribute to agglomeration of the crystalline regions of the cellulose. Acid hydrolyzed cCNCs showed greater crystallinity (69.0%) compared to either the rod-like (46.7%) or spherical (43.6%) cCNCs obtained from the sodium periodate method, which cleaves the chemical bond between the C2 and C3 positions, and retains regions of oxidized amorphous cellulose.

2.4. Class 4: Presence of Carboxylic Acid Groups at the Reducing End of CNCs

The presence of the highly reactive aldehyde functional group at the reducing end of CNCs allows for the production of cCNCs through another method. CNCs are made up of β-1,4 linked anhydro-D-glucose units. Similar to the structure of cellulose, they have three components: the non-reducing end, the cellobiose sequence (the repeat unit made up of two glucose units), and the reducing end. The latter is a hemiacetal, which in its open-chain form contains an aldehyde functional group. As such, these aldehyde moieties can be selectively oxidized to produce carboxyl groups at the reducing ends of the nanocrystals. Hieta et al. successfully applied a method involving the use of sodium chlorite at a pH of 3.5 (adjusted using acetic acid) for 20 h at room temperature to produce microfibrils and nanocrystals with carboxyl groups at the reducing ends of the materials. The selective deposition of silver at the reducing ends served as a tool to elucidate mechanistic and structural aspects related to the parallel configuration of cellulose I [63]. This method was later reproduced to yield thiolated CNCs using thioethanolamine via an amide linkage [64]. The aldehyde groups at the reducing ends were then oxidized to produce the carboxylic acid derivative using sodium chlorite (Scheme 7). The CNCs used were from sulfuric acid hydrolysis and were therefore devoid of surface carboxyl groups prior to treatment with sodium chlorite. This type of cCNCs is very useful for selective functionalization at the end of the nanocrystals.

$NaClO_2$

Scheme 7. Oxidation of aldehyde groups at the reducing ends of CNCs using sodium chlorite [63].

2.5. *cCNCs from Different Feedstocks*

Although commercial production of CNCs primarily uses cellulose-rich sources such as wood pulp, many researchers have turned to other feedstock materials that are either more regionally plentiful, or there is an imminent need to valorize specific biomass waste streams that would only be discarded into the environment if unused (such as agricultural residues). A key challenge to using any new biomass material is the weight percent content of cellulose versus its other components (i.e., lignin, hemicellulose, wax, and pectin) can vary from feedstock to feedstock. Table 1 summarizes some of the cCNCs previously discussed in Section 2, produced from different feedstock materials with a diverse range of physicochemical properties dependent on the method used. For the majority of the entries in Table 1, the feedstocks used to produce the cCNCs had undergone bleach processing prior to hydrolysis. Only the reactions utilizing APS were capable of hydrolyzing raw biomass into cCNCs.

Table 1. Production of cCNCs from different feedstock materials.

Feedstock	Method	Length (nm)	Width (nm)	CrI (%)	Yield (%)	Carboxyl Content (mmol/g)	Ref.
Cotton linters	HCl, TEMPO	n.r.	4–5	75	81	0.15 (DO)	[22]
Sugar beet pulp	HCl, TEMPO	n.r.	3–4	52	63	0.23 (DO)	[22]
Softwood-bleached kraft pulp	TEMPO, cavitation	155–244	3.6	81	94	1.57	[24]
MCC	TEMPO, cavitation	163–192	4.3	88	70	1.12	[24]
Flax fibers	APS	144 ± 5	3.8 ± 0.1	75	28	0.18 (DO)	[20]
Flax shives	APS	296 ± 16	5.1 ± 0.1	64	22	0.19 (DO)	[20]
Hemp fibers	APS	148 ± 3	5.8 ± 0.1	73	36	0.17 (DO)	[20]
Triticale straw extract	APS	134 ± 5	4.2 ± 0.1	73	31	0.11 (DO)	[20]
MCC	APS	128 ± 4	5.5 ± 0.1	83	65	0.19 (DO)	[20]
Whatman CF1	APS	121 ± 3	6.7 ± 0.3	91	81	0.12 (DO)	[20]
Wood pulp	APS	124 ± 6	6.0 ± 0.2	81	36	0.19 (DO)	[20]

Table 1. *Cont.*

Feedstock	Method	Length (nm)	Width (nm)	CrI (%)	Yield (%)	Carboxyl Content (mmol/g)	Ref.
Bacterial cellulose	APS	88 ± 5	6.5 ± 0.2	70	14	0.14 (DO)	[20]
Black spruce dissolving pulp	H_2O_2	170	10	n.r.	n.r.	0.15	[21,65]
Dissolving pulp	NaOCl, transition metal catalyzed	100–150	9–14	80	n.r.	0.15	[49]
Lemon seed	APS	140–160	10–20	74	13	n.r.	[41]
Lemon seed	TEMPO	340–380	26–42	66	52	n.r.	[41]
Juncus plant	Citric acid/HCl	352 ± 79	6.1 ± 2.8	83	n.r.	n.r.	[58]
Tomato plant residues	Citric acid/HCl	514 ± 131	4.7 ± 1.4	78	n.r.	n.r.	[59]
Bamboo borer powder	APS	20–50 (spheres)	-	63	n.r.	n.r.	[42]
Recycled paper	APS	130 ± 15	4.7 ± 2.5	72	22	0.57	[40]
Tunicate	APS	383 ± 164	10.9 ± 2.3	86	n.r.	0.07 (DO)	[44]
Cotton powder	APS, microwave	80–400	7	n.r.	46	n.r.	[39]
Softwood pulp	TEMPO	103 ± 29	21 ± 7	n.r.	12	0.7	[66]
Softwood pulp	Cu_2SO_4/H_2O_2	247 ± 58	6 ± 2	n.r.	51	0.9	[66]
Cotton pulp	APS/TMEDA	80–350	3–12	91	63	1.38	[37]
Cotton pulp	H_2SO_4, $KMnO_4$/oxalic acid	150–300	10–22	89	68	1.58	[51]
MCC	H_2SO_4, HNO_3	186 ± 13	9 ± 3	91	24	0.11	[52]
Sugarcane bagasse	Citric acid, ultrasonication	251 ± 52	21 ± 6	77	32	0.60	[55]
MCC	Citric acid, HCl	231.8 ± 23.2	15.8 ± 3.0	83	88	1.40	[56]
Bleached eucalyptus kraft pulp	Citric acid, $FeCl_3$	100–250	6–12	80	80	0.90	[60]
Bleached eucalyptus kraft pulp	Oxalic acid	278	19	81	25	0.19	[53]
Softwood pulp	Sodium periodate, sodium chlorite	120–200	13	91	n.r.	6.6	[61]
Bleached kraft pulp	H_2SO_4, sodium chlorite, 4% hypochlorite	150–200	5–20	n.r	50	n.r	[67]
Elephant grass	TEMPO, ultrasonication	174 ± 73	5.2 ± 1.4	n.r	n.r	n.r.	[32,33]
Bleached kraft pulp	$LiBr/H_2SO_4$, APS	57 ± 24	9.3 ± 3.1	96	52	0.44	[45]
Recycled medium-density fiberboard	APS	170–365	13–17	63	55	0.10 (DO)	[46]
Denim waste	NaOH, APS	81.5 ± 20.5	18.4 ± 2.5	27	79	n.r.	[47]
Balsa fibers	APS	95.5	6.5	n.r.	57.7	n.r.	[48]
Kapok fibers	APS	112.1	5.8	n.r.	60.7	n.r.	[48]

n.r. = not reported; DO: degree of oxidation.

3. Chemical Modifications of cCNCs

The presence of carboxyl groups on the surface of the nanocrystals not only provides electrostatic stabilization but also opens up a wide range of surface modification possibilities. It is possible to modify the surfaces of cCNCs through covalent and noncovalent functionalization. Compounds containing amino and hydroxyl groups can react covalently with the carboxyl moieties of the nanocrystals to form a wide range of derivatives. Similarly, it is possible to decorate the periphery of the nanocrystals with positively charged ions, polymers, or surfactants using noncovalent surface modifications, which rely on the electrostatic interactions with the negatively charged carboxylate groups. These modifications broaden the scope of applications of these nanomaterials by altering the surface properties and improving their dispersibility in non-aqueous media since unmodified cCNCs have limited dispersion in a non-aqueous environment, leading to flocculation.

3.1. Amidation

The most commonly used covalent reaction that cCNCs undergoes is the amidation reaction, which relies on the strong reactivity of the carboxyl and amino groups to produce the stable amide bond (Scheme 8). Amidation allows the incorporation of important functional moieties on the surface of the nanocrystals via the amide linkage that can be exploited for various applications. The most common conditions involve the method of Bulpitt and Aeschlimann [68], with the use of *N*-(3-dimethylaminopropyl)-N′-ethylcarbodiimide hydrochloride (EDC) and N-hydroxysuccimide (NHS), whereby EDC activates the carboxylic acid to form an O-acylisourea or carboxylic ester. Since the O-acylisourea is prone to hydrolysis, the addition of NHS creates a more stable NHS-ester that is resistant to hydrolysis. As the amine is added, the amidation reaction proceeds with less undesired products since NHS is a better-leaving group [69].

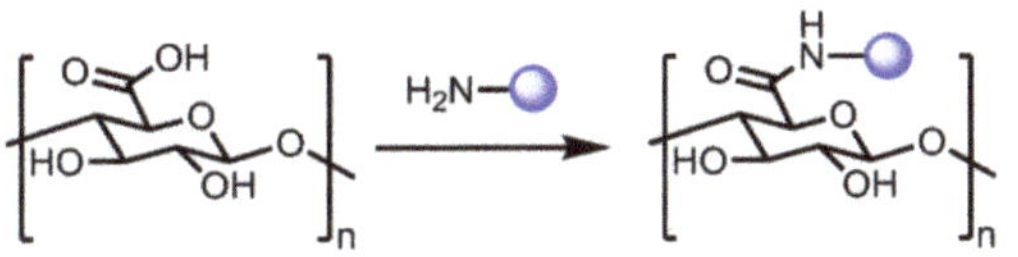

Scheme 8. General amidation reaction of cCNCs.

Table 2 shows examples of amidation reactions performed on cCNCs, with TEMPO oxidation being the most common way to generate carboxyl groups on the surface of the nanocrystals. An amidation reaction was performed on TEMPO-oxidized cCNCs from cotton linters and sugar beet pulp cellulose using 4-amino TEMPO in the presence of EDC and NHS at neutral pH [69]. The nitroxide moiety on the product was instrumental in characterizing the product through various spectroscopic techniques. It was found that about 30% of the carboxylic groups engaged in the newly formed amide bonds, with a slightly better coupling (31% compared to 26%) for the cotton linter due to a larger crystal size. The 4-amino-TEMPO-modified cCNCs formed stable dispersions in organic solvents such as chloroform, toluene, and dimethylformamide. Amidation at the reactive carboxyl acid sites was explored to generate precursors for click chemistry [70]. TEMPO-oxidized cCNCs were first modified with 11-azido-3,6,9-trioxaundecan-1-amine through a carbodiimide-mediated reaction to produce terminal azide functionalized CNCs. In a separate reaction, the amino-functional group on propargylamine was coupled with the cCNCs to yield CNCs with a terminal alkyne functionality. The resulting terminal azide and alkyne CNCs derivatives were then coupled via a Cu(I)-catalyzed Huisgen 1,3-dipolar cycloaddition to produce nanoplatelet gels, which were characterized by various spectroscopy and microscopy techniques. Using a similar strategy, the authors designed photoresponsive CNCs for the click reactions by simply using azido-bearing coumarin and anthracene as their azide precursors [71] to couple the propargylamine-CNC adduct formed [70]. Hemraz et al. reported a mild two-step protecting group-free protocol for the synthesis of aminated CNCs starting from sCNCs [72]. The latter was first oxidized using TEMPO oxidation, after which the TEMPO-oxidized cCNCs were reacted with aliphatic diamines under aqueous conditions to produce CNCs functionalized with terminal amino groups. This approach of functionalizing with bifunctional amines of small alkyl length can serve as a handle for bioconjugation and has been recently used for polymer attachment [73]. The cCNCs have also been coupled to aromatic amines [74], resulting in aromatic amides. Due to the hydrophobic nature of these derivatives, these nanomaterials can be dispersed in organic solvents. Amidation reactions have also been performed on cCNCs generated using the APS process to functionalize the nanocrystals with decylamine, dioctylamine, 2-aminoanthracene, and avidin [20].

Table 2. Amidation reactions on cCNCs for the design of functionalized CNCs in various applications.

H_2N–	Reaction Conditions	Dispersion	DO of Ccnc [a]	Application	Reference
	EDC, NHS, pH 7.5–8 24 h, 50 °C	$CHCl_3$, THF, toluene	0.21	Nanocomposites	[69]
	EDC, NHS, MES pH 4, 24 h, RT	Ionic liquid	0.2–0.28	Click chemistry Nanoplatelet gels	[70]
	EDC, NHS, MES pH 4, 24 h, RT	Ionic liquid	0.2–0.28	Click chemistry Nanoplatelet gels	[70,71]
	EDC, NHS, MES pH 5, 24 h, RT	Aqueous	0.25	Conjugation	[72]
	EDC, NHS, pH 8–8.5 72 h, RT	Acetone CH_2Cl_2	0.22	Nanocomposites	[74]
	EDC, NHS, pH 7.5–8 24 h, RT	DMF/aqueous	0.2	Thermo-responsive materials	[75]
	EDC, NHS, pH 7.5–8 48 h, RT	Aqueous	0.25	Thermo-responsive materials	[76]
	EDC, NHS, pH 7.2, PBS, DMF16 h, RT	DMF/aqueous	1.52 [b]	Nanosensors	[77]
Quantum Dots-NH_2	EDC, NHS, pH 5, overnight, RT	Aqueous	0.24	Bioimaging	[78]
Amino-functionalized sugars	EDC, NHS, MES pH 7.5–8.5 16 h, RT	Aqueous	1.56 [b]	Biorecognition	[79]
Fluorescent quinoline dye	EDC, NHS, MES pH 7.5–8.5 16 h, RT	Aqueous	1.56 [b]	Bioimaging	[79]
Aminoethyl rhodamine dye	CDI, DMF, 20 h, 60 °C	DMF/aqueous	1.102 [b]	Molecular switching, imaging	[80]
Lysozyme-NH_2	EDC, NHS, MES pH 5.7, 3 h, RT	Aqueous	-	Antibacterial	[81]
Lysozyme-NH_2	DMTMM, 1 h, 37 °C	Aqueous	0.7–0.9 [b]	Emulsifier	[66]

[a] The values displayed for the DO or carboxyl content of cCNC correspond to values prior to the amidation reactions; [b] Carboxyl content (mmol/g); EDC: *N*-(3-dimethylaminopropyl)-N′-ethylcarbodiimide hydrochloride; NHS: N-hydroxysuccimide; MES: 2-(N-morpholino)-ethanesulfonic acid buffer; DMTMM: 4-(4,6-dimethoxy-1,3,5-tr,azin-2-yl)-4-methylmorpholinium chloride; CDI: 1,1′-carbonyldiimidazole.

3.2. Esterification

Esterification is a common reaction performed on CNCs due to the abundance of primary alcohols on the surface, resulting in the reaction of hydroxyl groups from CNCs with carboxyl acids [10]. The presence of the carboxylic acid group on the surface of cCNCs still allows the scope for esterification via coupling with alcohols. This approach was used to design smart biobased materials for food packaging using cyclodextrins [82]. cCNCs were first prepared through a TEMPO-mediated oxidation reaction. The resulting product was then cast into a sheet and esterified with betacyclodextrin (β-CD) and hydroxypropyl-beta-cyclodextrin (HP β-CD) at 70 °C (Scheme 9). The formation of the ester bonds was confirmed by FT-IR, which showed adsorption bands at around 1730 cm^{-1} and the CD content was determined by phenolphthalein colorimetry. The small molecules, carvacrol and curcumin, were then entrapped by the β-CD and HP β-CD modified cCNCs and their release from the complexes was investigated. It was found that the loading amount of carvacrol and curcumin was significantly increased by the presence of CDs when compared to the complexation of the small molecules with the cCNCs. A prolonged release of carvacrol and curcumin was also observed. In addition, the carvacrol-loaded HP β-CD modified cCNCs displayed excellent antibacterial activities and could have applications as antibacterial products in packaging.

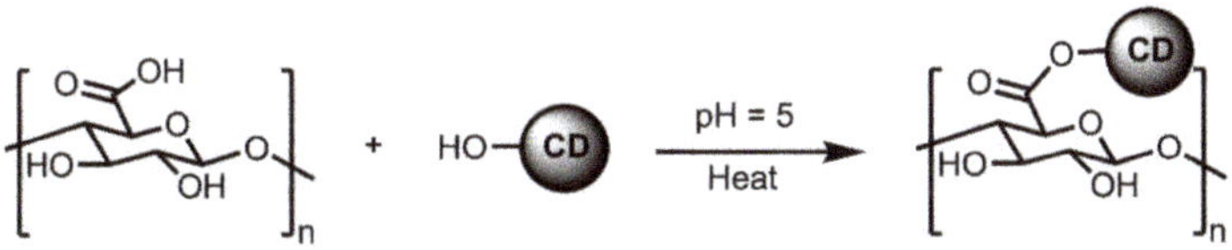

Scheme 9. Esterification reaction between cyclodextrins and TEMPO-oxidized cCNCs [82].

4. Applications of cCNCs

The number of applications utilizing cCNCs as an alternative material over the predominant sCNCs has increased in the past decade as the methods for cCNCs' production mature (Section 2), and researchers have begun to take advantage of both the inherent physicochemical properties and the potential to further tailor the material by surface modification at the carboxyl group (Section 3). Some of these applications will be discussed in detail. Naturally, as a reinforcing material, cCNCs have been used in polymer composites. However, with the biocompatibility of CNC, some of these applications have been directed towards the food and medical sector. The rich surface chemistry and nanoscale structure also enable cCNC to be a structural support for the fabrication of novel chemical and biocatalysts and materials for environmental remediation.

4.1. Nanocomposites

CNCs have been used to form nanocomposites due to numerous properties, namely, nanometer dimensions, low toxicity, biocompatibility, biodegradability, high aspect ratio, high Young's modulus, high strength, renewability, and abundance. Ever since the use of CNCs in polymer matrices was reported in 1995, there have been numerous reports of CNCs in nanocomposites [6,83]. This has also led to the application of cCNCs as components for nanocomposites, specifically utilizing the carboxyl group of the cCNCs to add functional groups that impart specific properties to the whole composite. For example, cCNCs have been used to create organic-inorganic nanomaterials through the carbodiimide-assisted coupling of the polyhedral oligomeric silsesquioxane (POSS) onto the carboxylic acid functional groups. POSS was found to be evenly distributed on the TEMPO-oxidized cCNCs, consistent with covalent bonding along the oxidized surface of C6 groups. However, for sodium periodate-oxidized cCNCs, POSS was found aggregated at different sites of the cellulose chain at either the more crystalline, central regions of the cellulose (where the carboxylic acid groups are located on the C2 and C3) or the oxidized amorphous end chains. Not surprisingly, the thermal stability of all the organic–

inorganic nanomaterials increased compared to the starting cCNCs as POSS can impart flame-retardant properties to composites [84].

Thermo-responsive polymers such as Jeffamines [75] and PNIPAM [76] have been grafted onto the surface of CNCs. These polymers can either be hydrophilic below the LCST or hydrophobic above the LCST just by changing conformation of the polymer chains from coil to globule state. Jeffamines are commercial polymers formed from the statistical copolymerization of ethylene oxide and propylene oxide that possess thermosensitive properties. Grafting of thermosensitive amine-terminated Jeffamine polymer chains on the surface of TEMPO-oxidized cCNCs was achieved using a "grafting-onto" strategy, and the resulting materials were characterized by various spectroscopy and microscopy techniques. The loss in electrostatic stability was reflected by the zeta potential measurements. In water, the zeta potential of the TEMPO-oxidized cCNCs was −23.5 mV, and upon grafting, as expected, the potential values decreased in magnitude in the range of −8 to −5 mV. Yet, despite losing on electrostatic stability, the samples did not flocculate, as they gained steric stability from the polymer chains. Through DLS experiments, which showed an increase in hydrodynamic diameter as the temperature was increased, the UV-vis spectra showed a surge in optical density above the LCST. This change in turbidity was reversible with a 4 °C hysteresis effect [75]. While in the case of sCNCs, surface modification was conducted via different polymerization approaches through the primary hydroxyl sites, attachment of PNIPAM chains in this work occurred through the covalent bonds formed between the amino groups of PNIPAM and the carboxylic acid groups of the cCNCs [76]. The TEMPO-oxidized cCNCs had a DO of 0.25, and the percentage of carboxylate groups that were found to have been substituted by amino groups was 30%. The removal of the unreacted polymer chains was ensured by quenching the reaction at a pH of 1–2 and subjecting the resulting product to dialysis. It was found that the polymer chains in PNIPAM-CNC bestowed steric stabilization in not only aqueous suspensions but also in suspensions of high ionic strengths resulting from the addition of salt, thus preventing flocculation. DLS experiments were used to demonstrate a thermo-reversible aggregation, whereby a low hydrodynamic diameter was observed below the LCST. Above the LCST, aggregation led to a five-fold increase in size. In addition to the thermo-sensitive behavior observed by rheological measurements, an increase in viscosity was observed from 0.008 for the TEMPO-oxidized cCNCs to 40 Pa·s for the grafted polymer. This change in viscosity was significantly higher than what was observed for PNIPAM adsorbed on the surface of the nanocrystals (0.3 Pa·s).

Risteen et al. designed a thermo-responsive poly(N-isopropylacrylamide) (PNIPAM)-based CNC derivative with a thermally switchable liquid-crystalline (LC) phase [85]. PNIPAM has been explored for a wide range of applications, including applications in the biomedical field, since it exhibits a lower critical solution temperature (LCST) at 32 °C. Usually, CNCs with a non-selective attachment of PNIPAM grafts undergo phase separation when heated above the LCST. In this work, the authors conducted a preferential attachment of an atom transfer radical polymerization (ATRP) initiator to the ends of the CNC rods, followed by a surface-initiated ATRP to produce a selectively grafted CNC sample (Scheme 10). The authors used CNCs from USDA, which already contain some carboxyl groups on the surface (26 μmol COOH/g CNCs). The concentration of reducing end groups in the unmodified CNCs was found to be 18.2 μmol CHO per gram of CNCs. The aldehyde functional group at the reducing ends of the CNCs was oxidized using sodium chlorite to produce the carboxylated material at a yield of 57%. The resulting carboxylic acids were then reacted with ethylenediamine to express primary amino groups on the surface of the CNCs. A ninhydrin assay was conducted to determine the concentration of primary amine groups introduced, and it was found to be much higher than expected (60.3 μmol NH_2/g CNCs). The ethylenediamine molecules did not just covalently react to the carboxylic acids on the surface and ends but also physically adsorbed on the nanocrystals. The aminated CNCs were then subjected to polymerization to produce birefringent materials. Unlike the "brush" PNIPAM-modified CNCs, which have densely packed polymer chains, these selectively grafted CNC have limited polymer chains on the surface, allowing more translational and rotational freedom with minimal attractive interactions. As a result, heating and cooling did not result in phase separation for the selectively grafted CNCs. A similar approach was used to graft thermosensitive polyetheramines at the reducing ends of CNCs [86]. Heating the polymer-functionalized CNC dispersions above the LCST triggered a heat-induced aggregation, leading to the formation of star-shaped assemblies. As the temperature was lowered below the LCST, the assemblies dissociated. This temperature-dependent aggregation and dissociation was reversible and could be monitored by DLS experiments through a change in hydrodynamic diameter. The authors envisage the use of these thermal-responsive CNCs for sensor applications.

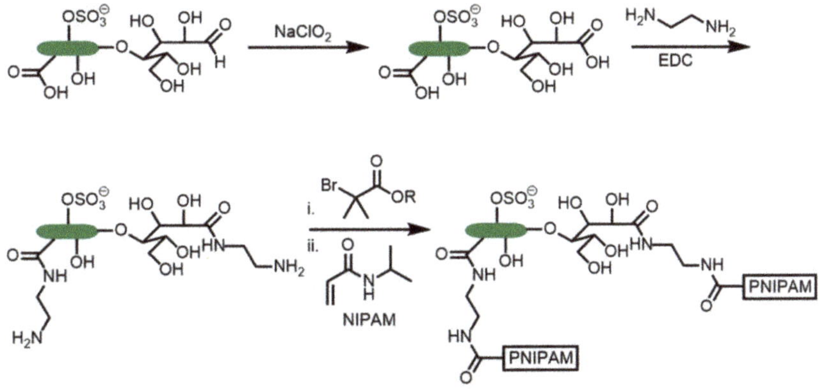

Scheme 10. PNIPAM functionalized cCNCs, with preferential attachment at reducing ends of cCNCs [85].

Capron and co-workers investigated the modification of bacterial cellulose nanocrystals (BCNs) for the formation of silver nanoparticles [87]. Silver nanoparticles are popular amendments to CNCs as they can confer antimicrobial properties [88]. BCNs were obtained from nata de coco, a jelly-like food produced from the fermentation of coconut water in which *Gluconacetobacter xylinu* secretes microbial cellulose. BCNs exhibit primarily hydroxyl groups, and the researchers modified these nanocrystals to be either cationic (covalent bonding with cholaminchloride hydrochloride), anionic (by TEMPO oxidation), or hydrophobic (EDC-NHS coupling of TEMPO-oxidized cCNCs to octylamine). Native BCNs and the modified BCNs had lengths between 850 and 1750 nm and a width between 20 and 40 nm. The surface charge of native BCNs (−11.2 mV) decreased to −25.4 mV upon TEMPO-oxidation, increased to +15.9 mV in cationic modification, and remained similar in charge when hydrophobically modified (−9.3 mV). Contrary to the assumption that increased negatively charged surfaces would preferentially favor the interaction of Ag^+ for nanoparticle formation over positively charged surfaces, the key finding was that hydroxyl surface groups on cellulosic surfaces were the real nanoparticle nucleation points for well-defined Ag nanoparticles in all cases. Additional negative surface charges merely improved the dispersion state, thereby increasing the accessibility to the nucleation sites. The ability to form well-dispersed Ag nanoparticles on even hydrophobic BCNs demonstrates the potential to form bifunctional nanocomposite materials in non-aqueous solvents.

The use of self-healing conductive composites with metallic conductivity has been proposed for flexible electronics for enhanced durability and operational lifetime, as these materials are able to restore their structure and functionality from damage [89]. As a biocompatible and hydrophilic polymer, polyvinylalcohol (PVA) composites have been used for electronic skin sensors but suffer from poor mechanical properties, which limits the stretchability of the material [90]. Chen et al. turned to cCNC as a nanofiller to enhance the mechanical properties of flexible films for personal electronics [91]. cCNCs obtained from the hydrolysis of MCC with citric acid were initially coated with polypyrrole (PPy) to form a cCNC-Ppy conductive nano-network, followed by further polymerization with methyl methacrylate (MMA) onto the functionalized CNCs. Ppy is a commonly used, low-cost, conductive polymer, promising conductive polymers with fast impulse discharge performance and vitro/vivo cytocompatibility for personal electronics [92]. PVA was incorporated at different concentrations (1–4%) to the cCNC-Ppy-PMMA composite to form the final conductive, self-healing films characterized by a dynamic network crosslinked by hydrogen bonding. As expected, the addition of cCNCs increased the mechanical strength of the films. However, variations in the concentration of PVA (and their effect on the overall hydrogen-bonding network) also influenced the tensile strength, fracture stress, and fracture strain of the films. As the films were stretched, bent, or deformed, an electric resistance response was obtained, which correlated to the distance of the PPy chains. The self-healing properties of these films were demonstrated by cutting the film and then reheating the cut ends together at 60 °C for 20 min to show that a continuous circuit could be completed to light an LED bulb.

Food packaging has an integral role in the preservation of perishable foods and allows for the optimal use of space during food transport and storage. Most food packaging materials are synthetic polymers that are cheap to produce and offer the required durability and waterproofness required for food preservation. Unfortunately, these polymers are extremely resistant to natural degradation, which leads to significant environmental pollution. There has been an interest identifying biopolymers that can offer similar physicochemical properties to plastics yet are biodegradable and safe for food use. Cao et al. initially developed cassia gum films for edible packaging films assisted by the use of glycerol and sorbitol as plasticizing agents [93]. Cassia gum is a water-soluble polysaccharide extracted from the seeds of the sicklepod consisting of 1,4-ß-D-mannopyranose units and 1,6-linked α-D-galactopyranose units. In these films, they noted their poor mechanical, heat-sealability, and barrier properties. Nanoscale materials have been extensively investigated as potential filler materials in composites for strength reinforcement and impart functionalities, and cCNCs (with dimensions of 100–500 nm in length and 3–10 nm in width) were used in conjunction with glycerol to produce reinforced cassia gum films [94]. The interaction between the cassia gum and cCNCs was based on hydrogen bonding, with no evidence of covalent interactions between the two respective polysaccharides. The inclusion of up to 4% cCNCs to these films resulted in improving the oil permeability, mechanical properties, and heat seal strength compared to the native cassia gum film. At 6% cCNC content, aggregation of the cassia gum and cCNCs was observed.

4.2. Biomedical Applications

CNCs are promising materials for biomedical applications due to their hydrophilicity, biocompatibility, their high specific surface area, high aspect ratio, and sharp angles. They have been explored in drug delivery and tissue engineering applications as versatile platforms for protein and enzyme immobilization and in combination with fluorescent probes for bioimaging and biosensing [14]. A fluorescent CNC probe was designed as a potential candidate for nanosensor applications using a 1,8-naphthalimide dye, covalently attached to TEMPO-oxidized cCNCs via an EDC/NHS mediated coupling between the carboxylic acid and amino groups [77]. The resulting reaction was confirmed by FT-IR, UV-vis adsorption, and fluorescence spectra. In addition, no morphological or structural change was observed with the surface-functionalized nanofibers. Since the fluorescence quantum efficiency of dye depends on solvent polarity, the effect of solvent polarity and ionic strength on fluorescent intensity was investigated. It was found that the stronger fluorescence was observed under UV illumination for the dye-labeled CNCs compared to the pure dye. In addition, the colour was enhanced at lower solvent permittivity and higher ionic strength. This behaviour was attributed to aggregation-enhanced emission, resulting from the compression of the electrostatic double layer of the nanocrystals.

Chu et al. went a step further and developed fluorescent nanoprobes by conjugating the amino groups of another 1,8-naphthalimide dye (NANI) and biocompatible poly (ethylene glycol) (PEG) to cCNCs, obtained from the TEMPO/HCl hydrolysis of bleached softwood pulp, through two-step successive grafting [95]. Notably, the addition of PEG brushes on the cCNCs increased the hydrophilicity of the nanomaterial. The resulting materials were investigated in the bioimaging of Hela cells in a physiological environment at high salt concentration. The nanometer dimensions and the morphology were vital in allowing the CNC derivatives to enter and disperse in the cells efficiently. As a result, strong fluorescence emission was obtained in the bioimaging. Nanoprobes consisting of only cCNCs and NANI were found to aggregate within the cell, leading to lower fluorescence emission, but nanoprobes with cCNCs, NANI, and PEG were found to be better dispersed within the cell, leading to greater fluorescence emission.

Rojas and co-workers coupled amino-functionalized carbon quantum dots (CQD) with TEMPO-oxidized cCNCs to produce biocompatible and photoluminescent hybrid material, which was used as a bioimaging probe to image HeLa and RAW 264.7 macrophage cells [78]. The resulting material was characterized using X-ray photoelectron spectroscopy, fluorescence spectroscopy, electron, and confocal microscopies. Unlike the cCNC, the CQD-cCNC hybrid material displayed a green fluorescence upon excitation using UV light (365 nm). It was found that the quantum dots improved the cytocompatibility and internalization of the cCNCs in the cells.

A multifunctional CNC platform constituting carbohydrate and fluorophore components was developed for lectin recognition and bacterial imaging [79]. The carboxylic acid groups of the TEMPO-oxidized cCNCs were coupled with the amino groups on the fluorescent dye 4-(2-aminoethylamino)-7H-benz[de]benzimidazo[2,1-a]isoquinoline-7-one and the carbohydrate ligand 1-(2-(2-(2-aminoethoxy)ethoxy)ethoxy-D-mannopyranoside via the amide linkage, in the presence of EDC and NHS (Scheme 11). As evidenced by scanning transmission electron microscopy (STEM) images, these dual-functionalized CNCs maintained a rod-like morphology, with an average length of 265 $\pm$ 80 nm and width of 5.2 $\pm$ 0.3 nm. They also displayed a CrI of 65% and a yellow-green fluorescence upon excitation at 450 nm. DLS was used to investigate the change in the apparent size due to aggregation from lectin binding. A significant increase in hydrodynamic diameter was observed from about 200 nm for the cCNCs to around 400–600 nm, which could be attributed to cross-linking between the lectins and the carbohydrate-functionalized CNCs. These results were supported by transmission electron microscopy (TEM) measurements, which showed agglomeration of the modified CNCs and the protein concanavalin A. Confocal fluorescence microscopy was used to evaluate the biorecognition and binding of the dual-functionalized CNCs to concanavalin A through interactions with the mannose moiety and it was found that mannosylated nanocrystals underwent selective interactions with FimH-presenting *E. coli*.

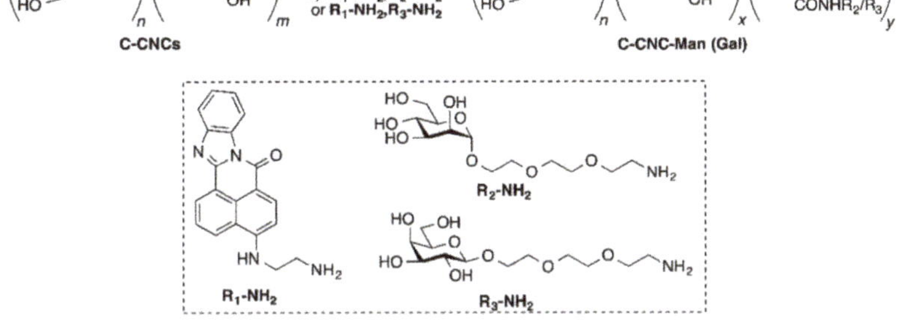

1). EDC·HCl, NHS

2). R_1-NH_2, R_2-NH_2 or R_1-NH_2, R_3-NH_2

C-CNCs

C-CNC-Man (Gal)

R_1-NH_2

R_2-NH_2

R_3-NH_2

Scheme 11. Synthesis of carbohydrate-functionalized and fluorescent cCNCs (Reproduced with permission from [79], Copyright: American Chemical Society 2015).

Zhao et al. immobilized rhodamine spiroamide groups on the surface of CNCs in an attempt to design smart materials [80]. The rhodamine B dye was first coupled with ethylenediamine to introduce amino groups to the dye. The resulting aminoethyl rhodamine was then reacted with TEMPO-oxidized cCNCs by coupling the carboxylate groups of the nanoparticles to the primary amino groups on the dye. Elemental analysis was used to determine the amount of aminoethyl rhodamine immobilized on the surface of the nanocrystals, and it was estimated to be about 0.2 mmol/g. These functionalized materials (CNC-RhB) were responsive towards multiple stimuli and underwent a color-switching process as a result of a ring-opening and -closing process initiated by the various triggers applied. As seen in Scheme 12, CNC-RhB displayed a beige color in DMF at neutral pH and room temperature, and subsequent heat treatment at 130 °C led to a ring-opening reaction upon which the color transitioned to yellow. The ring-opening mechanism of the rhodamine B is reversible with the addition of NaOH, which restores the temporary binding between carboxyl groups and rhodamine. Similarly, UV-illumination under acidic conditions produced a magenta-colored solution. These materials have potential applications as smart switchable devices due to their excellent switching performance and reversibility under multiple stimuli. It was also possible to study the structural changes by nuclear magnetic resonance through dynamic nuclear polarization.

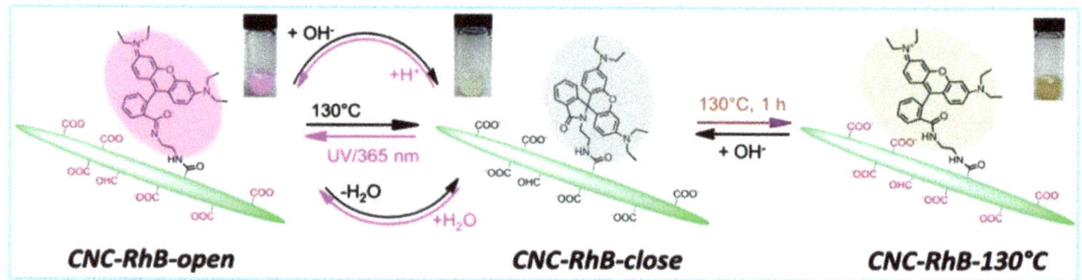

Scheme 12. Synthesis of rhodamine-functionalized fluorescent cCNCs (Reproduced with permission from [80], Copyright: Royal Society of Chemistry 2014).

Biomaterials have attracted considerable interest in tissue-engineering applications due to their hydrophilic properties and biocompatibility to mimic components of the extracellular matrix. Challenges in the development of artificial extracellular matrix materials include adequate engagement of cell attachment, proliferation, migration, and differentiation, and their ability to transport biomolecules and waste metabolites from native tissue. Recently, extrusion 3D printing of cell-free or cell-loaded hydrogel ink has led to the production of desired compositions and architectures for tissue-engineering applications [96]. However, most biomaterials on their own lack the inherent mechanical properties required to maintain their structural integrity and support stress factors in an in vivo 3D environment [97]. Kumar et al. developed stable cell-free and cell-loaded hydrogel inks for direct-write extrusion-based 3D printing using cCNCs and xanthan gum within a sodium alginate hydrogel matrix [98]. On their own, each biopolymer component lacks the necessary physicochemical properties to produce sufficient bioinks for 3D printing. Sodium alginate has fast gelling properties but inadequate mechanical properties and biocompatibility to interact with the extracellular matrix. Xanthan gum features highly pseudoplastic behavior, good thermal properties, and variable viscosity influenced by shear rate, but poor mechanical properties. The cCNCs (prepared from MCC using APS) were selected specifically for both their reinforcement properties as a nanofiller, and to access their carboxylic acid functional groups for extra crosslinking to the hydrogel network for improving the mechanical properties and shear-thinning behavior during extrusion. Utilizing all three biopolymers, the bioink featured good rheological properties, post-printing fidelity, and dynamic mechanical properties of the gel with viability towards human skin fibroblast (CCD-986Sk) cells.

Lastly, CNCs have been applied towards the formation of Pickering emulsions [99], as they can self-assemble at the oil-water interface due to negative surface charges, wettability, crystalline structure, and morphology. Pickering emulsions can be utilized in a number of biomedical applications, including drug delivery and scaffolding material [100]. Their properties have been Capron and co-workers investigated the ability of sCNCs to form Pickering emulsions and found that samples with the higher negative surface charges produced more stable emulsions [101]. Unlike the fairly labile sulfate half-esters produced during the hydrolysis process, the presence of carboxylate groups from oxidation of the CNCs increased the magnitude and stability of surface charges on the crystals. Surface modification of the oxidized nanocrystals through the ionic exchange of the sodium ions with stearyltrimethylammonium chloride produced CNCs decorated with quaternary ammonium salts, which could form more stable emulsions and larger droplets. Mikulcová et al. investigated APS-oxidized CNCs in the formation of Pickering emulsions from acidic to neutral pH with a triglyceride oil (e.g., tricaprine or tricapryline) [102]. The size of emulsion droplets was dependent on oil and cCNC loading. The authors noted that lowering the pH did not trigger the release of oil from the micron-sized Pickering emulsions, which they attributed to the strong absorption of cCNCs towards the polar triglyceride oil used.

4.3. Biocatalysis

In biocatalysis, enzymes are often immobilized on scaffold materials to improve their stability and increase their activity over a free and unsupported enzyme. Nanoscale carriers are favored because of their high surface area, which can increase enzyme loading and reduce diffusion restrictions, leading to enhanced enzyme activity while promoting greater stability over a free enzyme.

Abouhmad et al. investigated the immobilization of lysozymes on the surface of CNCs and its subsequent impact on antibacterial activity [81]. Lysozyme is an antibacterial enzyme that is effective against Gram positive bacteria, while T4 bacteriophage shows activity against both Gram positive and Gram negative bacteria. Lysozyme, from hen egg white and T4 bacteriophage, were covalently linked to cCNCs and glutaraldehyde-activated aminated CNCs. The latter was prepared by first forming the ammonium salt of cCNCs through neutralization of cCNCs with ammonium hydroxide to improve dispersibility. The neutralized cCNCs were then oxidized to their dialdehyde derivative with sodium metaperiodate, followed by reaction with ethylenediamine and subsequent activation by glutaraldehyde. Incubation of lysozymes with the glutaraldehyde-activated aminated CNCs led to conjugates showed increased antibacterial activity and exhibited enhanced storage stability. The effect of immobilization on antibacterial activity and enzymatic activity was also investigated, and it was found that lysozymes adsorbed on CNCs showed weak activities compared to the lysozyme-CNC bioconjugates formed through covalent immobilization. The activities exhibited by the materials are likely related to their surface properties, as evidenced by the zeta potential measurements. The cCNCs, ammonium salt of cCNCs, and glutaraldehyde-activated aminated CNCs showed zeta potential values of about −54 mV, −42 mV, and +31 mV, respectively, while the zeta potentials of the lysozymes were mildly positive (about +7 to +10 mV). Noncovalent adsorption of the lysozymes onto the CNC surface resulted in negatively charged bioconjugates (about −30 mV), while the covalently bound glutaraldehyde-CNC-lysozyme conjugates formed stable and positively charged colloidal suspensions, with zeta potential values ranging between +37 mV and +43 mV.

Koshani and van den Ven also developed a nanocarrier system using cCNCs for the immobilization of lysozymes [66]. cCNCs were targeted as a potential nanocarrier for enzymes due to their non-toxicity, ease of fabrication and modification, biodegradability, and biocompatibility. The nanomaterials were produced from softwood pulp under two methods: TEMPO-oxidation and Cu-catalyzed H_2O_2 oxidation [103]. The carboxyl content of H_2O_2-oxidized cCNCs (0.9 mmol/g) was higher than TEMPO-oxidized cCNCs

(0.7 mmol/g). The carboxyl groups of H_2O_2-oxidized cCNCs enabled the immobilization of the lysozymes by adsorption to achieve an enzyme loading of 240 mg/g by adsorption for an immobilization yield of 73% and an enzyme activity of 62%. In comparison, TEMPO-oxidized cCNCs exhibited an enzyme loading of 550 mg/g by adsorption for an immobilization yield of 98% and an enzyme activity of 73%. When a bioconjugation method employing DMTMM was used to immobilize the lysozyme, an inverse relation was obtained in which the H_2O_2-oxidized cCNCs obtained an immobilization yield of 65% and an enzyme activity of 60%, compared to an immobilization yield of 54% and an enzyme activity of 52%. As expected, the lower carboxyl content of TEMPO-oxidized CNCs should result in lower enzyme immobilization due to the necessity for carboxylic groups for bio-conjugation. On the other hand, for adsorption, the authors attribute the differences in enzyme immobilization and activity to changes in the structural conformation of lysozymes, influenced by the varying surface roughness and curvature of the different cCNCs.

4.4. Catalysis

The high surface area and available surface functional groups have made CNCs an ideal support material for recyclable catalysts. The most developed approaches for CNC-catalysts are as support materials for metal nanoparticles as they can promote the reduction of metals, but other possibilities include the grafting of organometallic species or surface functionalization to create organocatalysts [104]. The surface chemistry of CNCs can promote greater interaction of substrates with the supported catalytic site for reactions to occur more favorably. The majority of catalyst examples utilize sCNCs; however, a few examples for cCNCs have been reported in the literature. An early example of using APS-oxidized cCNCs in a catalyst application was reported by Luong and co-workers for the reduction of 4-nitrophenol to 4-aminophenol [105]. Gold nanoparticles are often used as catalysts for organic transformation reactions, but an agglomeration of these catalysts in a solution can significantly reduce their catalytic activity. Although gold nanoparticles can be stabilized by immobilization on other support materials, these immobilized catalyses often exhibit weaker catalytic activity compared to the free metal catalyst due to lower accessibility of substrates towards immobilized gold nanoparticles. A nanocomposite was prepared by deposition of pre-formed negatively charged gold nanoparticles (AuNP) onto the surface of a positively charged PDDA-coated cCNCs in which its catalytic activity of AuNP/PDDA/cCNCs was demonstrated in the reduction 4-nitrophenol to 4-aminophenol. Tam and co-workers developed a polyamidoamine (PAMAM) dendrimer-grafted cCNCs (G6 PAMAM) as a support material for gold nanoparticles [106]. PAMAM dendrimers were grafted onto TEMPO-oxidized cCNCs followed by in situ reductions of $HAuCl_4$ to form gold nanoparticles with diameters of <20 nm without the need for further addition of a reducing agent. Gold nanoparticle formation was influenced by pH and the concentration of the PAMAM-grafted cCNCs: larger nanoparticles tend to form at pH > 3.3 and at higher concentrations of the functionalized cCNCs. The developed catalyst materials were applied towards the reduction of 4-nitrophenol to 4-aminophenol in which its enhanced catalytic behavior was attributed to smaller gold nanoparticles, and the improved dispersity and accessibility of gold nanoparticles within the PAMAM dendrimer domain. Other reports of nanocomposites made of PANAM dendrimers grafted onto TEMPO-oxidized cCNCs have been applied for the removal of Cu (II) from water solutions with the highest Cu (II) adsorption capacity of 92.07 mg/g at 25 °C [73], and for CO_2 capture at capacities of 13.31 ± 0.38 mg/g at 25 °C, 9.64 ± 0.60 mg/g at 35 °C, and 9.18 ± 1.27 mg/g at 45 °C, respectively [107].

The carboxyl groups of cCNCs offer sites of reactivity for the covalent immobilization of catalytic transition metal complexes that would normally be soluble and difficult to recover in the reaction media. Heterogeneous dirhodium (II) catalysts based on cCNCs were produced for cycloproponation reactions [108]. Ligand exchange between $Rh_2(OOCCF_3)_4$ and the carboxyl groups was conducted to incorporate the dirhodium (II) catalyst to the surface of the TEMPO-oxidized cCNCs. Using thermogravimetric analysis, the dirhodium (II) content on the cCNC surface was found to be 0.23 mmol/g of cCNCs. The catalytic performance of Rh grafted-cCNC catalyst was investigated in the cyclopropanation of styrene with ethyl diazoacetate, achieving a yield of 80% in 180 min at room temperature compared to the homogeneous catalyst ($Rh_2(OOCCF_3)_4$) of 84% in 1 min. Not surprisingly, the authors believe the activity of heterogeneous catalysts was limited by mass transfer and loss of mobility as the Rh catalytic centers are confined on the surface of the cCNC support. However, the benefits of the heterogeneous catalyst reside in its stability against leaching and its reusability associated with the covalent bonding of the Rh to the cCNC and catalyst accessibility from separate binding sites on the surface of the nanocrystal.

Hu et al. immobilized reactive Co (III) salen species onto sCNCs to produce heterogeneous catalysts for the synthesis of cyclic carbonates from propylene oxide and CO_2 [109]. These catalysts achieved a near quantitative yield in 24 h at low CO_2 pressure of 0.1 MPa with demonstrated reusability in four reaction cycles. Control experiments showed a 22% yield in the absence of any catalyst, but when sCNCs and TEMPO-oxidized cCNCs were used (without the presence of Co (III) in both cases), the yields increased to 35% and 55%, respectively. The authors believe that the surface-functional groups (sulfate or carboxylic groups) could promote catalytic activity to the system through hydrogen bonding between the epoxide starting material and the CNC.

Ellebrecht and Jones investigated the use of cCNC as a support material for bifunctional heterogeneous organocatalysts [110]. Alkyl diamines were conjugated onto TEMPO-oxidized cCNCs (derived initially from sCNCs) by EDC amide coupling to impart both acid and base characteristics to cCNC surfaces. The catalyst materials were used as acid–base catalysts for the aldol condensation reaction of 4-nitrobenzaldehyde with acetone, with comparable activity to aminosilica organocatalysts. Greater rates of reactions were achieved when the solvent system was switched from a water–acetone mixture to an all-organic system (acetonitrile–acetone) system. The authors also noted that the length of diamines used was influential in decreasing catalytic activity: longer diamines promoted crosslinking between the cCNCs while shorter diamines reduced acid–base cooperativity in the overall organocatalyst system. The authors optimized their work on these acid–base heterogeneous catalysts by specifically tailoring surface species on the cCNCs [111]. The sulfate half-ester content was found to not affect catalysis but was essential to organocatalyst functionalization and dispersion. Crosslinking reduces catalyst activity, and diaminopropane was found to be the most optimal diamine base as this chain length led to the least number of crosslinking events between functionalized cCNCs. By controlling the carboxylic acid to amine ratio, the authors were also able to prepare cCNC catalysts effective for the upgrading of furfural (a key specialty chemical derived from biomass resources) by aldol condensation with acetone with high selectivity towards the furfuryl acetone, outperforming mesoporous silica SBA-15-supported catalysts in both activity and selectivity.

4.5. Environmental Remediation

The release of harmful organic pollutants in waste streams poses a major environmental and human health concern. Amongst various pollutants, organic dyes are very worrisome due to their complex aromatic structures, which make them chemically stable and resistant to degradation by heat, light, and oxidants in nature. One of the most cost-effective methods for dye remediation is adsorption. Pinto et al. examined the potential of two positively charged CNCs, namely sCNCs and cCNCs, for the removal of the organic cationic dye, Auramine O (AO) [112]. In head-to-head batch adsorption experiments, the highest removal percentage (82%) and adsorption capacity (20 mg/g) were obtained for the sCNCs for an equilibrium contact time of 30 min. Adsorption behavior of AO on both CNCs was well-described by Freundlich isotherm. The calculated thermodynamics parameters indicated that AO adsorption on both CNC samples was a spontaneous exothermic process. In contrast, while entropy decreased for the sCNCs, an increase in entropy was reported for cCNCs. Overall, the adsorption process of AO dye on the CNC was thought to be driven by two mechanisms: electrostatic interactions between the positively charged nitrogen of AO with anionic sulfate or carboxylate groups of the respective CNC, and hydrogen bonding between the AO tertiary nitrogen atoms with the CNC hydroxyl groups.

Samadder et al. developed magnetic composites based on polyacrylic acid (PAA) crosslinked with acrylic-functionalized Fe_3O_4 nanoparticles (M3D) and cCNCs for the removal of the widely used and toxic cationic dye, methylene blue, from aqueous solutions [113]. The cCNCs were extracted from sawdust waste. The latter first underwent stepwise treatments to remove lignin, pectin, and hemicellulose, and the resulting purified cellulose was subjected to sulfuric acid hydrolysis, followed by TEMPO oxidation. The nanopolysaccharide reduced the gel-like properties of the whole nanocomposite particles (60–90 nm in size), which was found to be advantageous for use as an adsorbent. The maximum adsorption capacity increased from 114 mg/g for the M3D–PAA nanocomposite to 332 mg/g for the M3D-PAA-cCNC nanoparticles for methylene blue. These nanoparticles displayed better performance in alkaline conditions as the surface charge of the particles became increasingly negative.

In another remediation report involving methylene blue, Luong and co-workers utilized the high aspect ratio of cCNCs, and the negative surface charges resulting from the carboxylate groups generated via the APS process, to adsorb the toxic cationic dye [114]. It was found that cCNC has a homogeneous surface and a maximum adsorption capacity of 101 mg per gram of nanomaterial, only moderately lower than the maximum adsorption capacity of 127 mg/g for a carbon monolith. Ethanol was 90% effective in desorbing methylene blue from cCNCs using centrifugation. Adsorption and remediation of the dye using cCNCs and ethanol could be a potential method for the removal of organic pollutants. It should be noted that sCNCs from cellulose fibers, as well as its carboxylated derivative obtained from subsequent TEMPO oxidation, also demonstrate adsorption abilities. While the TEMPO oxidation certainly adds an extra step towards the formation of the cCNCs, the adsorption capacity improved from 118 mg to 769 mg of methylene blue per gram for cCNCs [115].

Most of the studies conducted on the adsorption capacity of cCNCs relied on the high aspect ratio of the rod-like nanoparticles. A recent study provided new insights into the ability of cellulose microbeads, made from cCNCs by a hydrogen peroxide method, to adsorb methylene blue [65]. The dye was spray-dried onto the cCNC microbeads, and the adsorption was quantified using a film-pore diffusion model. This work highlights the transport properties of the microbeads and shows how nanoparticles can be rescaled into functional microscale objects for water purification and applications requiring controlled uptake and release.

The removal of salts such as NaCl, Na_2SO_4, and $MgSO_4$ is an important task with respect to water remediation, particularly for the treatment of brackish water desalination. Salt-rejecting membranes generally require high permeate flux and selectivity, as well as good mechanical properties to purify drinking water [116]. Nanomaterials have been used as additives in membranes to increase their performance and reduce environmental impact. The use of sCNCs as a beneficial addition to polyamide (PA)-polyethersulfone (PES) membranes for enhancing nanofiltration of Na_2SO_4 and NaCl aqueous solutions was investigated [117]. The sCNC layer was found to not only control the degree of cross-linking of the formed PA layer but also increased the hydrophilicity of the membranes, allowing for greater water permeation through CNC-containing membranes compared to PES and PA-PES membranes alone. The effects of where the TEMPO-oxidized cCNCs were incorporated in the production of thin-film membranes also consisting of PA and PES for water desalination of Na_2SO_4, $MgSO_4$, and NaCl was studied [118]. A higher permeate flux was reported when the cCNC was incorporated into the active layer of the membrane. However, when the cCNC was added to the support layer of the membrane, increased salt rejection and mechanical performance were achieved for the thin-film membranes.

Nanoabsorbents formed from the functionalization of sCNCs with succinic anhydride were prepared for the removal of heavy metal ions (Scheme 13) [119]. The resulting succinic carboxylated CNC (SCNC) was then converted into the sodium form by treatment with a saturated sodium bicarbonate solution and freeze-dried to yield the sodium nanoabsorbent (NaSCNC). Adsorption experiments conducted on SCNC and NaSCNC demonstrated that the nanomaterials were able to adsorb the heavy metal ions Cd^{2+} and Pb^{2+}. A higher adsorption capacity was exhibited by NaSCNC due to the ability of the carboxylate to bind more strongly to the nanomaterial than in the acid form. In addition, it was found that adsorption was pH-dependent, with higher adsorption ability for the metal ions at higher pH. The authors believe that at lower pH, more protons are competing with metal ions for the active binding sites. At the same time, the adsorbent surface is positively charged, which increases the difficulty for positively charged metal ions to approach the adsorbent due to electrostatic repulsion. The maximum adsorption capacities of SCNC and NaSCNC were found to be 259.7 mg/g and 344.8 mg/g for Cd^{2+}, and 367.6 mg/g and 465.1 mg/g for Pb^{2+}, respectively. It was interesting to note that both SCNC and NaSCNC showed high selectivity and binding for Pb^{2+}, and adsorption was not affected by the presence of coexisting ions. NaSCNC was also efficiently regenerated upon treatment with a mild saturated NaCl solution, with no loss of adsorption capacity after two recycles.

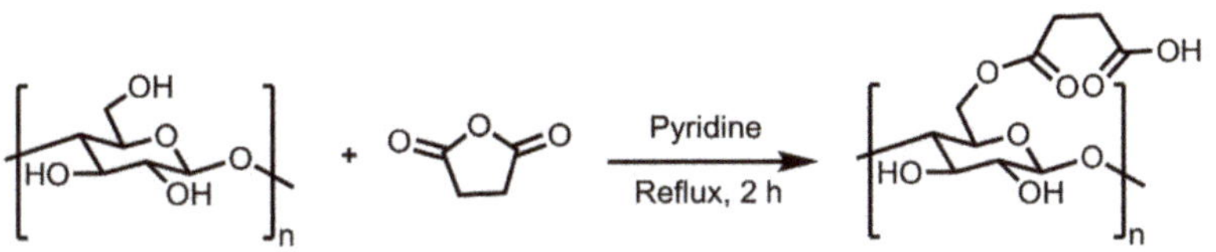

Pyridine
Reflux, 2 h

Scheme 13. Reaction of sCNCs with succinic anhydride to produce another type of carboxylated CNC [119].

Going a step further, Lu et al. designed a magnetic composite made of Fe_3O_4 nanoparticles and cCNCs (Fe_3O_4-CNC) for adsorption of Pb^{2+}, which did not require filtration or centrifugation to recover the adsorbent [57]. The dicarboxylated CNCs were produced using the oxidative hydrolysis of MCC by APS, followed by esterification with citric acid. With two carboxylic acid moieties per AGU (Scheme 5), the resulting product was co-precipitated with Fe_3O_4 nanoparticles to form a composite (Fe_3O_4-CNC) with enhanced dispersibility. With the templating effect of the cCNC, the hydroxyl and carboxylic acid groups on the cCNC surface hydrogen-bonded with the hydroxyl groups on the surface of the Fe_3O_4 nanoparticles. The Fe_3O_4 succeeded in imparting a lower but sufficiently high saturation magnetization for Fe_3O_4-CNC (34.13 versus 71.08 emu·g^{-1}) to enable fast magnetic separation. Adsorption experiments conducted on Fe_3O_4-CNC showed a maximum adsorption capacity of 63.78 mg/g for Pb^{2+} and easy separation from the aqueous solution using magnetism. Dicarboxylated CNCs also improved coagulation and flocculation performance for cationic dyes, kaolin suspension, and textile effluent [120]. The dicarboxylated CNCs, with an approximate dimension of 200–250 nm in length, were generated from a citric acid and HCl hydrolysis of MCC, with a reaction time of 2–6 h and yields of about 80–90%. Compared to the TEMPO-oxidized cCNCs [115], these dicarboxylated CNCs showed a higher adsorption capacity for methylene blue removal. In the presence of the coagulant $CaCl_2$, dicarboxylated CNCs displayed remarkable coagulation–flocculation potential by showing a turbidity removal of 99.5% for kaolin suspension. Similarly, dicarboxylated CNCs produced from hydrochloric acid and citric acid treatment of bamboo have also been used as templates to produce zinc oxide-CNC hybrid materials to adsorb within 5 min cationic dyes methylene blue and malachite green at about 91% and 98%, respectively [121]. In addition, the zinc oxide-CNC materials exhibited high antibacterial activities against *E. coli* and *S. aureus*.

4.6. Rheology Modifiers

Researchers have turned to cCNCs as novel ingredients for rheology modification with enhanced lubricating and heat-transfer properties. In tribology, a common challenge is to identify lubricants for moving systems that will work well over large temperature ranges and have low-temperature fluidity at nominal viscosity. Nanomaterials such as metal oxides and carbonaceous materials have attracted considerable attention as additives in lubricants due to their inherent properties to decrease wear and friction in moving objects caused by high pressures and temperatures [122]. These nanoparticles can interact better than larger particles to form surface protective films, which contribute to the improvement of anti-wear performances. However, metal oxides themselves are under scrutiny as they can contribute to corrosion and introduce impurities that are considered toxic. Awang et al. investigated cCNC as an additive for SAE40 engine oil and found that 0.1% cCNC exhibited the lowest coefficient of friction (COF) and the strongest wear resistance under all conditions in the piston skirt-liner tribometer test [122]. In the absence of cCNC, worn surfaces showed large areas of metal exfoliation, metal burr, and wear debris. However, with cCNCs in the engine oil, the cCNCs are able to chemically react with surfaces to form a tribo-boundary film that deposits above the frictional surfaces and lower the COF, resulting in minimal surface exfoliation.

The removal of heat is essential for maintaining the longevity of tools that would otherwise wear out and fail due to excessive heat generation. As an alternative to metal oxides, Samylingam et al. investigated the use of cCNC as a sustainable additive for coolant fluid for machining [123]. These fluids consist of cCNCs (ranging in concentration from 0.1 to 1.5%), ethylene glycol, and water. As the concentration of cCNCs increased, the coolant fluid exhibited greater thermal conductivity and viscosity. A coolant fluid of 0.5% cCNC improved tool life through superior heat transfer fluid in comparison to the benchmark metalworking fluid used for lathe machining operation.

Bentonite is an absorbent swelling clay consisting mostly of montmorillonite. It has attracted interest as suspensions for water-based drilling fluids in oil well excavation but suffers from poor rheology and filtration performance at low solid content. Li et al. compared the use of cCNC and its cationically modified derivative (caCNC-reaction of cCNC with 2,3-epoxypropyl)trimethylammonium chloride) as rheology and filtration modifiers for bentonite suspensions. cCNC (at 0.5% concentration) were found to be better rheological and filtration agents in water-based drilling fluids compared to caCNC [124]. This was directly attributed to how the bentonite suspensions dispersed when cCNCs and cationic cCNCs were added. For native cCNCs, the bentonite platelets were bridged together by the cCNCs in an "edge-to-edge" interaction (Scheme 14). Cationic cCNCs were absorbed on the face surface of bentonite platelets leading to stack layers of platelets in a "face-to-face" interaction. The authors believe the edge-to-edge interaction created by the cCNC favors a more rigid network, thus inducing superior rheological performance over caCNC. In the corresponding patent application, cCNCs were found to be superior to non-carboxylated CNCs as they provide a more uniform dispersion state of the bentonite platelets attributed to the interactions between the carboxyl groups on the nanocrystals and the weakly positively charged edge surfaces of the bentonite platelet [125].

The same group later developed water-based drilling fluids with thermo-controllable rheological properties in response to the need for stimuli-responsive drilling fluids that can maintain rheology based on the ever-changing conditions (i.e., temperature, pH, pressure, etc.) that oil reservoirs exist in. This was accomplished through cCNCs grafting with poly(2-acrylamido-2-methyl-1-propanesulfonic acid) (PAMPS) and PNIPAM by APS free radical polymerization [126]. PAMPS, with its amido functional groups and negatively charged sulfonate groups, promoted the creation of well-dispersed bentonite clusters, while PNIPAM enabled thermo-thickening rheological characteristics for the fluid at elevated temperatures.

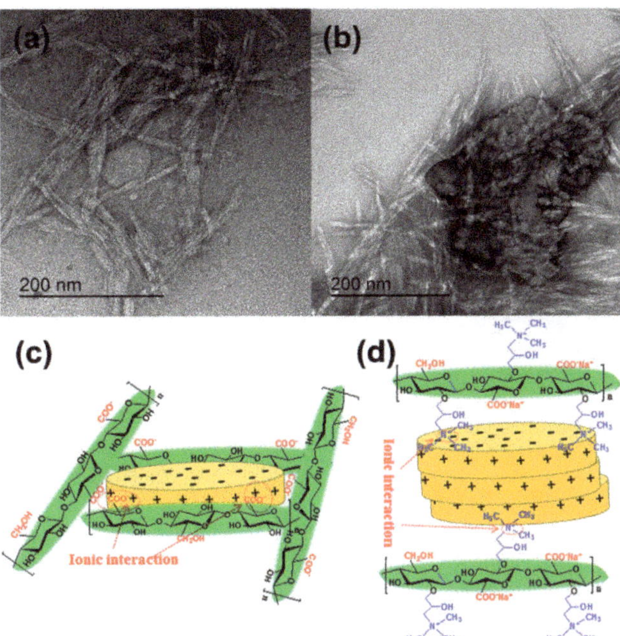

Scheme 14. TEM micrographs and proposed surface interactions of water-based drilling fluids formulated using cCNC and bentonite in an edge-to-edge interaction (**a**,**c**) and cationic CNC and bentonite in a face-to-face interaction (**b**,**d**). (Reproduced with permission from [124]. Copyright: American Chemical Society, 2018).

5. Conclusions and Future Outlook

In this review, we present the known production methods for generating cCNCs. These range from (1) post-production modification of CNCs by TEMPO oxidation, (2) direct conversion of biomass to cCNCs using oxidizers, (3) treatment of cellulosic sources with organic acids, to (4) transformation of cellulose-rich substrates to cCNCs using a mixture of oxidizers, organic acids, and mechanical processing. The presence of the carboxyl groups on the surface of cCNCs provides unique opportunities for the carboxylated material to supersede the predominant sCNCs. Indeed, compared to sCNCs, cCNCs can be easily modified through various chemical transformations at the carboxyl group, while still maintaining their dispersibility. Such surface modification is a gateway to incorporating new functional properties for the design of composites, hydrogels, Pickering emulsions, and films that can be applied to biomedical, agrochemical, and electronic domains. Other less-explored opportunities reside in the use of cCNCs as scaffold materials for enzymes and catalysis.

Large-scale utilization of sCNCs is only beginning to take place, which has necessitated the need for ton-scale production of sCNCs. The current cCNCs production is only at a pilot scale, and it is likely that cCNCs will face similar challenges in identifying anchor applications to justify upscaling the production of the nanomaterial. Feedstock availability, batch-to-batch consistency, and the reduction of water and chemical consumption are critical barriers to the commercial production of cCNCs that need to be overcome. Most current processes utilize an exorbitant amount of acids and oxidizers; therefore, efforts should be focused on making cCNCs under milder and greener reaction conditions. This could entail developing enzymes that could selectively attack the amorphous regions of cellulose and transform the primary hydroxyl groups into carboxyl groups. Enabling the recycling of reagents would also help in reducing the detrimental environmental impact of these oxidative processes. Finally, little is known on the environmental impact of CNCs as there is no information available on the fate of these entities in the environment nor on life-cycle analysis [127]. Efforts aiming at addressing these pertinent environmental questions are necessary for all types of CNCs.

Author Contributions: Conceptualization and writing—E.L. and U.D.H. All authors have read and agreed to the published version of the manuscript.

Funding: This work was funded internally by the National Research Council of Canada.

Acknowledgments: The authors would like to thank the National Research Council of Canada for supporting this work and Fanny Monteil-Rivera for her valuable input.

Conflicts of Interest: The authors declare no conflict of interest.

References

1. Nickerson, R.; Habrle, J. Cellulose intercrystalline structure. *Ind. Eng. Chem.* **1947**, *39*, 1507–1512. [CrossRef]
2. Wu, Q.; Meng, Y.; Wang, S.; Li, Y.; Fu, S.; Ma, L.; Harper, D. Rheological behavior of cellulose nanocrystal suspension: Influence of concentration and aspect ratio. *J. Appl. Polym. Sci.* **2014**, *131*. [CrossRef]
3. Viet, D.; Beck-Candanedo, S.; Gray, D.G. Dispersion of cellulose nanocrystals in polar organic solvents. *Cellulose* **2007**, *14*, 109–113. [CrossRef]
4. Revol, J.-F.; Bradford, H.; Giasson, J.; Marchessault, R.; Gray, D. Helicoidal self-ordering of cellulose microfibrils in aqueous suspension. *Int. J. Biol. Macromol.* **1992**, *14*, 170–172. [CrossRef]
5. Roman, M. Toxicity of cellulose nanocrystals: A review. *Ind. Biotechnol.* **2015**, *11*, 25–33. [CrossRef]
6. Favier, V.; Chanzy, H.; Cavaillé, J. Polymer nanocomposites reinforced by cellulose whiskers. *Macromolecules* **1995**, *28*, 6365–6367. [CrossRef]
7. Liew, S.Y.; Thielemans, W.; Walsh, D.A. Electrochemical capacitance of nanocomposite polypyrrole/cellulose films. *J. Phys. Chem. C* **2010**, *114*, 17926–17933. [CrossRef]
8. Jackson, J.K.; Letchford, K.; Wasserman, B.Z.; Ye, L.; Hamad, W.Y.; Burt, H.M. The use of nanocrystalline cellulose for the binding and controlled release of drugs. *Int. J. Nanomed.* **2011**, *6*, 321.
9. Frka-Petesic, B.; Guidetti, G.; Kamita, G.; Vignolini, S. Controlling the photonic properties of cholesteric cellulose nanocrystal films with magnets. *Adv. Mater.* **2017**, *29*, 1701469. [CrossRef]
10. Habibi, Y.; Lucia, L.A.; Rojas, O.J. Cellulose Nanocrystals: Chemistry, Self-Assembly, and Applications. *Chem. Rev.* **2010**, *110*, 3479–3500. [CrossRef]
11. Moon, R.J.; Martini, A.; Nairn, J.; Simonsen, J.; Youngblood, J. Cellulose nanomaterials review: Structure, properties and nanocomposites. *Chem. Soc. Rev.* **2011**, *40*. [CrossRef] [PubMed]
12. Klemm, D.; Kramer, F.; Moritz, S.; Lindström, T.; Ankerfors, M.; Gray, D.; Dorris, A. Nanocelluloses: A New Family of Nature-Based Materials. *Angew. Chem. Int. Ed.* **2011**, *50*, 5438–5466. [CrossRef] [PubMed]
13. Lam, E.; Male, K.B.; Chong, J.H.; Leung, A.C.W.; Luong, J.H.T. Applications of functionalized and nanoparticle-modified nanocrystalline cellulose. *Trends Biotechnol.* **2012**, *30*, 283–290. [CrossRef]
14. Sunasee, R.; Hemraz, U.D.; Ckless, K. Cellulose nanocrystals: A versatile nanoplatform for emerging biomedical applications. *Expert Opin. Drug Deliv.* **2016**, *13*, 1243–1256. [CrossRef]
15. Vanderfleet, O.M.; Cranston, E.D. Production routes to tailor the performance of cellulose nanocrystals. *Nat. Rev. Mater.* **2020**. [CrossRef]
16. Sunasee, R. Nanocellulose: Preparation, Functionalization and Applications. In *Reference Module in Chemistry, Molecular Sciences and Chemical Engineering*; Elsevier: Amsterdam, The Netherlands, 2020.
17. Lu, P.; Hsieh, Y.-L. Preparation and properties of cellulose nanocrystals: Rods, spheres, and network. *Carbohydr. Polym.* **2010**, *82*, 329–336. [CrossRef]
18. Mukherjee, S.; Woods, H. X-ray and electron microscope studies of the degradation of cellulose by sulphuric acid. *Biochim. Biophys. Acta* **1953**, *10*, 499–511. [CrossRef]
19. Filson, P.B.; Dawson-Andoh, B.E.; Schwegler-Berry, D. Enzymatic-mediated production of cellulose nanocrystals from recycled pulp. *Green Chem.* **2009**, *11*, 1808–1814. [CrossRef]
20. Leung, A.C.; Hrapovic, S.; Lam, E.; Liu, Y.; Male, K.B.; Mahmoud, K.A.; Luong, J.H. Characteristics and properties of carboxylated cellulose nanocrystals prepared from a novel one-step procedure. *Small* **2011**, *7*, 302–305. [CrossRef]
21. Andrews, M.P.; Morse, T. Method for producing functionalized nanocrystalline cellulose and functionalized nanocrystalline cellulose thereby produced. Patent EP3174905A1, 27 November 2011.
22. Montanari, S.; Roumani, M.; Heux, L.; Vignon, M.R. Topochemistry of carboxylated cellulose nanocrystals resulting from TEMPO-mediated oxidation. *Macromolecules* **2005**, *38*, 1665–1671. [CrossRef]
23. Kang, X.; Kuga, S.; Wang, C.; Zhao, Y.; Wu, M.; Huang, Y. Green preparation of cellulose nanocrystal and its application. *Acs Sustain. Chem. Eng.* **2018**, *6*, 2954–2960. [CrossRef]

24. Zhou, Y.; Saito, T.; Bergström, L.; Isogai, A. Acid-free preparation of cellulose nanocrystals by TEMPO oxidation and subsequent cavitation. *Biomacromolecules* **2018**, *19*, 633–639. [CrossRef] [PubMed]
25. Qin, Z.-Y.; Tong, G.; Chin, Y.F.; Zhou, J.-C. Preparation of ultrasonic-assisted high carboxylate content cellulose nanocrystals by TEMPO oxidation. *BioResources* **2011**, *6*, 1136–1146.
26. Vanderfleet, O.M.; Reid, M.S.; Bras, J.; Heux, L.; Godoy-Vargas, J.; Panga, M.K.; Cranston, E.D. Insight into thermal stability of cellulose nanocrystals from new hydrolysis methods with acid blends. *Cellulose* **2019**, *26*, 507–528. [CrossRef]
27. Cherhal, F.; Cousin, F.; Capron, I. Influence of charge density and ionic strength on the aggregation process of cellulose nanocrystals in aqueous suspension, as revealed by small-angle neutron scattering. *Langmuir* **2015**, *31*, 5596–5602. [CrossRef]
28. Yu, H.; Qin, Z.; Liang, B.; Liu, N.; Zhou, Z.; Chen, L. Facile extraction of thermally stable cellulose nanocrystals with a high yield of 93% through hydrochloric acid hydrolysis under hydrothermal conditions. *J. Mater. Chem. A* **2013**, *1*, 3938–3944. [CrossRef]
29. Camarero Espinosa, S.; Kuhnt, T.; Foster, E.J.; Weder, C. Isolation of thermally stable cellulose nanocrystals by phosphoric acid hydrolysis. *Biomacromolecules* **2013**, *14*, 1223–1230. [CrossRef] [PubMed]
30. De Nooy, A.E.J.; Besemer, A.C.; van Bekkum, H. Highly selective tempo mediated oxidation of primary alcohol groups in polysaccharides. *Recl. Des. Trav. Chim. Des. Pays-Bas* **1994**, *113*, 165–166. [CrossRef]
31. Habibi, Y.; Chanzy, H.; Vignon, M.R. TEMPO-mediated surface oxidation of cellulose whiskers. *Cellulose* **2006**, *13*, 679–687. [CrossRef]
32. Noronha, V.T.; Camargos, C.H.M.; Jackson, J.C.; Souza Filho, A.G.; Paula, A.J.; Rezende, C.A.; Faria, A.F. Physical Membrane-Stress-Mediated Antimicrobial Properties of Cellulose Nanocrystals. *ACS Sustain. Chem. Eng.* **2021**, *9*, 3203–3212. [CrossRef]
33. Jackson, J.C.; Camargos, C.H.M.; Noronha, V.T.; Paula, A.J.; Rezende, C.A.; Faria, A.F. Sustainable Cellulose Nanocrystals for Improved Antimicrobial Properties of Thin Film Composite Membranes. *ACS Sustain. Chem. Eng.* **2021**, *9*, 6534–6540. [CrossRef]
34. Salajková, M.; Berglund, L.A.; Zhou, Q. Hydrophobic cellulose nanocrystals modified with quaternary ammonium salts. *J. Mater. Chem.* **2012**, *22*, 19798–19805. [CrossRef]
35. Saito, T.; Nishiyama, Y.; Putaux, J.-L.; Vignon, M.; Isogai, A. Homogeneous Suspensions of Individualized Microfibrils from TEMPO-Catalyzed Oxidation of Native Cellulose. *Biomacromolecules* **2006**, *7*, 1687–1691. [CrossRef] [PubMed]
36. Lam, E.; Leung, A.C.; Liu, Y.; Majid, E.; Hrapovic, S.; Male, K.B.; Luong, J.H. Green strategy guided by Raman spectroscopy for the synthesis of ammonium carboxylated nanocrystalline cellulose and the recovery of byproducts. *ACS Sustain. Chem. Eng.* **2013**, *1*, 278–283. [CrossRef]
37. Liu, Y.; Liu, L.; Wang, K.; Zhang, H.; Yuan, Y.; Wei, H.; Wang, X.; Duan, Y.; Zhou, L.; Zhang, J. Modified ammonium persulfate oxidations for efficient preparation of carboxylated cellulose nanocrystals. *Carbohydr. Polym.* **2020**, *229*, 115572. [CrossRef]
38. Feng, X.D.; Guo, X.Q.; Qiu, K.Y. Study of the initiation mechanism of the vinyl polymerization with the system persulfate/N, *N*, *N′*, *N′*-tetramethylethylenediamine. *Die Makromol. Chem. Macromol. Chem. Phys.* **1988**, *189*, 77–83. [CrossRef]
39. Amoroso, L.; Muratore, G.; Ortenzi, M.A.; Gazzotti, S.; Limbo, S.; Piergiovanni, L. Fast Production of Cellulose Nanocrystals by Hydrolytic-Oxidative Microwave-Assisted Treatment. *Polymers* **2020**, *12*, 68. [CrossRef]
40. Jiang, Q.; Xing, X.; Jing, Y.; Han, Y. Preparation of cellulose nanocrystals based on waste paper via different systems. *Int. J. Biol. Macromol.* **2020**, *149*, 1318–1322. [CrossRef]
41. Zhang, H.; Chen, Y.; Wang, S.; Ma, L.; Yu, Y.; Dai, H.; Zhang, Y. Extraction and comparison of cellulose nanocrystals from lemon (*Citrus limon*) seeds using sulfuric acid hydrolysis and oxidation methods. *Carbohydr. Polym.* **2020**, *238*, 116180. [CrossRef]
42. Hu, Y.; Tang, L.; Lu, Q.; Wang, S.; Chen, X.; Huang, B. Preparation of cellulose nanocrystals and carboxylated cellulose nanocrystals from borer powder of bamboo. *Cellulose* **2014**, *21*, 1611–1618. [CrossRef]
43. Lu, P.; Hsieh, Y.-L. Cellulose isolation and core–shell nanostructures of cellulose nanocrystals from chardonnay grape skins. *Carbohydr. Polym.* **2012**, *87*, 2546–2553. [CrossRef]
44. He, J.; Bian, K.; Piao, G. Self-assembly properties of carboxylated tunicate cellulose nanocrystals prepared by ammonium persulfate oxidation and subsequent ultrasonication. *Carbohydr. Polym.* **2020**, *249*, 116835. [CrossRef] [PubMed]
45. Li, N.; Bian, H.; Zhu, J.Y.; Ciesielski, P.N.; Pan, X. Tailorable cellulose II nanocrystals (CNC II) prepared in mildly acidic lithium bromide trihydrate (MALBTH). *Green Chem.* **2021**, *23*, 2778–2791. [CrossRef]
46. Khanjanzadeh, H.; Park, B.-D. Optimum oxidation for direct and efficient extraction of carboxylated cellulose nanocrystals from recycled MDF fibers by ammonium persulfate. *Carbohydr. Polym.* **2021**, *251*. [CrossRef] [PubMed]
47. Culsum, N.T.U.; Melinda, C.; Leman, I.; Wibowo, A.; Budhi, Y.W. Isolation and characterization of cellulose nanocrystals (CNCs) from industrial denim waste using ammonium persulfate. *Mater. Today Commun.* **2021**, *26*. [CrossRef]
48. Marwanto, M.; Maulana, M.I.; Febrianto, F.; Wistara, N.J.; Nikmatin, S.; Masruchin, N.; Zaini, L.H.; Lee, S.-H.; Kim, N.-H. Effect of Oxidation Time on the Properties of Cellulose Nanocrystals Prepared from Balsa and Kapok Fibers Using Ammonium Persulfate. *Polymers* **2021**, *13*, 1894. [CrossRef]
49. Mcalpine, S.; Nakoneshny, J. Production of crystalline cellulose. Patent WO2017127938A1, 8 March 2017.
50. Available online: https://umaine.edu/pdc/nanocellulose/2021/ (accessed on 24 May 2021).
51. Zhou, L.; Li, N.; Shu, J.; Liu, Y.; Wang, K.; Cui, X.; Yuan, Y.; Ding, B.; Geng, Y.; Wang, Z.; et al. One-Pot Preparation of Carboxylated Cellulose Nanocrystals and Their Liquid Crystalline Behaviors. *ACS Sustain. Chem. Eng.* **2018**, *6*, 12403–12410. [CrossRef]
52. Cheng, M.; Qin, Z.; Hu, J.; Liu, Q.; Wei, T.; Li, W.; Ling, Y.; Liu, B. Facile and rapid one–step extraction of carboxylated cellulose nanocrystals by H2SO4/HNO3 mixed acid hydrolysis. *Carbohydr. Polym.* **2020**, *231*, 115701. [CrossRef]

53. Chen, L.; Zhu, J.; Baez, C.; Kitin, P.; Elder, T. Highly thermal-stable and functional cellulose nanocrystals and nanofibrils produced using fully recyclable organic acids. *Green Chem.* **2016**, *18*, 3835–3843. [CrossRef]
54. Li, B.; Xu, W.; Kronlund, D.; Määttänen, A.; Liu, J.; Smått, J.-H.; Peltonen, J.; Willför, S.; Mu, X.; Xu, C. Cellulose nanocrystals prepared via formic acid hydrolysis followed by TEMPO-mediated oxidation. *Carbohydr. Polym.* **2015**, *133*, 605–612. [CrossRef]
55. Ji, H.; Xiang, Z.; Qi, H.; Han, T.; Pranovich, A.; Song, T. Strategy towards one-step preparation of carboxylic cellulose nanocrystals and nanofibrils with high yield, carboxylation and highly stable dispersibility using innocuous citric acid. *Green Chem.* **2019**, *21*, 1956–1964. [CrossRef]
56. Yu, H.; Abdalkarim, S.Y.H.; Zhang, H.; Wang, C.; Tam, K.C. Simple Process to Produce High-Yield Cellulose Nanocrystals Using Recyclable Citric/Hydrochloric Acids. *ACS Sustain. Chem. Eng.* **2019**, *7*, 4912–4923. [CrossRef]
57. Lu, J.; Jin, R.-N.; Liu, C.; Wang, Y.-F.; Ouyang, X.-k. Magnetic carboxylated cellulose nanocrystals as adsorbent for the removal of Pb(II) from aqueous solution. *Int. J. Biol. Macromol.* **2016**, *93*, 547–556. [CrossRef]
58. Kassab, Z.; Syafri, E.; Tamraoui, Y.; Hannache, H.; Qaiss, A.E.K.; El Achaby, M. Characteristics of sulfated and carboxylated cellulose nanocrystals extracted from Juncus plant stems. *Int. J. Biol. Macromol.* **2020**, *154*, 1419–1425. [CrossRef]
59. Kassab, Z.; Kassem, I.; Hannache, H.; Bouhfid, R.; El Achaby, M. Tomato plant residue as new renewable source for cellulose production: Extraction of cellulose nanocrystals with different surface functionalities. *Cellulose* **2020**, *27*, 1–17. [CrossRef]
60. Liu, W.; Du, H.; Liu, H.; Xie, H.; Xu, T.; Zhao, X.; Liu, Y.; Zhang, X.; Si, C. Highly Efficient and Sustainable Preparation of Carboxylic and Thermostable Cellulose Nanocrystals via FeCl3-Catalyzed Innocuous Citric Acid Hydrolysis. *Acs Sustain. Chem. Eng.* **2020**, *8*, 16691–16700. [CrossRef]
61. Yang, H.; Alam, M.N.; van de Ven, T.G.M. Highly charged nanocrystalline cellulose and dicarboxylated cellulose from periodate and chlorite oxidized cellulose fibers. *Cellulose* **2013**, *20*, 1865–1875. [CrossRef]
62. Huang, B.; He, H.; Liu, H.; Zhang, Y.; Peng, X.; Wang, B. Multi-type cellulose nanocrystals from sugarcane bagasse and their nanohybrids constructed with polyhedral oligomeric silsesquioxane. *Carbohydr. Polym.* **2020**, *227*, 115368. [CrossRef]
63. Hieta, K.; Kuga, S.; Usuda, M. Electron staining of reducing ends evidences a parallel-chain structure in Valonia cellulose. *Biopolymers* **1984**, *23*, 1807–1810. [CrossRef]
64. Arcot, L.R.; Lundahl, M.; Rojas, O.J.; Laine, J. Asymmetric cellulose nanocrystals: Thiolation of reducing end groups via NHS–EDC coupling. *Cellulose* **2014**, *21*, 4209–4218. [CrossRef]
65. Wu, J.; Andrews, M.P. Carboxylated Cellulose Nanocrystal Microbeads for Removal of Organic Dyes from Wastewater: Effects of Kinetics and Diffusion on Binding and Release. *ACS Appl. Nano Mater.* **2020**, *3*, 11217–11228. [CrossRef]
66. Koshani, R.; van de Ven, T.G. Carboxylated Cellulose Nanocrystals Developed by Cu-Assisted H_2O_2 Oxidation as Green Nanocarriers for Efficient Lysozyme Immobilization. *J. Agric. Food Chem.* **2020**, *68*, 5938–5950. [CrossRef] [PubMed]
67. Reiner, R.S.; Rudie, A.W. Process Scale-Up of Cellulose Nanocrystal Production to 25 kg per Batch at the Forest Products Laboratory. In *Production and Applications of Cellulose Nanomaterial*; Postek, M.T., Moon, R.J., Rudie, A.W., Bilodeau, M.A., Eds.; TAPPI: Atlanta, GA, USA, 2013; pp. 21–24.
68. Bulpitt, P.; Aeschlimann, D. New strategy for chemical modification of hyaluronic acid: Preparation of functionalized derivatives and their use in the formation of novel biocompatible hydrogels. *J. Biomed. Mater. Res.* **1999**, *47*, 152–169. [CrossRef]
69. Follain, N.; Marais, M.-F.; Montanari, S.; Vignon, M.R. Coupling onto surface carboxylated cellulose nanocrystals. *Polymer* **2010**, *51*, 5332–5344. [CrossRef]
70. Filpponen, I.; Argyropoulos, D.S. Regular Linking of Cellulose Nanocrystals via Click Chemistry: Synthesis and Formation of Cellulose Nanoplatelet Gels. *Biomacromolecules* **2010**, *11*, 1060–1066. [CrossRef] [PubMed]
71. Filpponen, I.; Sadeghifar, H.; Argyropoulos, D.S. Photoresponsive Cellulose Nanocrystals. *Nanomater. Nanotechnol.* **2011**, *1*. [CrossRef]
72. Hemraz, U.D.; Boluk, Y.; Sunasee, R. Amine-decorated nanocrystalline cellulose surfaces: Synthesis, characterization, and surface properties. *Can. J. Chem.* **2013**, *91*, 974–981. [CrossRef]
73. Wang, Y.; Lu, Q. Dendrimer functionalized nanocrystalline cellulose for Cu(II) removal. *Cellulose* **2019**, *27*, 2173–2187. [CrossRef]
74. Le Gars, M.; Delvart, A.; Roger, P.; Belgacem, M.N.; Bras, J. Amidation of TEMPO-oxidized cellulose nanocrystals using aromatic aminated molecules. *Colloid Polym. Sci.* **2020**, *298*, 603–617. [CrossRef]
75. Azzam, F.; Heux, L.; Putaux, J.-L.; Jean, B. Preparation by Grafting Onto, Characterization, and Properties of Thermally Responsive Polymer-Decorated Cellulose Nanocrystals. *Biomacromolecules* **2010**, *11*, 3652–3659. [CrossRef] [PubMed]
76. Gicquel, E.; Martin, C.; Heux, L.; Jean, B.; Bras, J. Adsorption versus grafting of poly(*N*-Isopropylacrylamide) in aqueous conditions on the surface of cellulose nanocrystals. *Carbohydr. Polym.* **2019**, *210*, 100–109. [CrossRef]
77. Wu, W.; Song, R.; Xu, Z.; Jing, Y.; Dai, H.; Fang, G. Fluorescent cellulose nanocrystals with responsiveness to solvent polarity and ionic strength. *Sens. Actuators B Chem.* **2018**, *275*, 490–498. [CrossRef]
78. Guo, J.; Liu, D.; Filpponen, I.; Johansson, L.-S.; Malho, J.-M.; Quraishi, S.; Liebner, F.; Santos, H.A.; Rojas, O.J. Photoluminescent Hybrids of Cellulose Nanocrystals and Carbon Quantum Dots as Cytocompatible Probes for in Vitro Bioimaging. *Biomacromolecules* **2017**, *18*, 2045–2055. [CrossRef]
79. Zhou, J.; Butchosa, N.; Jayawardena, H.S.; Park, J.; Zhou, Q.; Yan, M.; Ramstrom, O. Synthesis of multifunctional cellulose nanocrystals for lectin recognition and bacterial imaging. *Biomacromolecules* **2015**, *16*, 1426–1432. [CrossRef]

80. Zhao, L.; Li, W.; Plog, A.; Xu, Y.; Buntkowsky, G.; Gutmann, T.; Zhang, K. Multi-responsive cellulose nanocrystal–rhodamine conjugates: An advanced structure study by solid-state dynamic nuclear polarization (DNP) NMR. *Phys. Chem. Chem. Phys.* **2014**, *16*, 26322–26329. [CrossRef]
81. Abouhmad, A.; Dishisha, T.; Amin, M.A.; Hatti-Kaul, R. Immobilization to Positively Charged Cellulose Nanocrystals Enhances the Antibacterial Activity and Stability of Hen Egg White and T4 Lysozyme. *Biomacromolecules* **2017**, *18*, 1600–1608. [CrossRef]
82. De Castro, D.O.; Tabary, N.; Martel, B.; Gandini, A.; Belgacem, N.; Bras, J. Controlled release of carvacrol and curcumin: Bio-based food packaging by synergism action of TEMPO-oxidized cellulose nanocrystals and cyclodextrin. *Cellulose* **2018**, *25*, 1249–1263. [CrossRef]
83. Favier, V.; Canova, G.R.; Cavaillé, J.Y.; Chanzy, H.; Dufresne, A.; Gauthier, C. Nanocomposite materials from latex and cellulose whiskers. *Polym. Adv. Technol.* **1995**, *6*, 351–355. [CrossRef]
84. Cordes, D.B.; Lickiss, P.D.; Rataboul, F. Recent Developments in the Chemistry of Cubic Polyhedral Oligosilsesquioxanes. *Chem. Rev.* **2010**, *110*, 2081–2173. [CrossRef]
85. Risteen, B.; Delepierre, G.; Srinivasarao, M.; Weder, C.; Russo, P.; Reichmanis, E.; Zoppe, J. Thermally Switchable Liquid Crystals Based on Cellulose Nanocrystals with Patchy Polymer Grafts. *Small* **2018**, *14*. [CrossRef]
86. Lin, F.; Cousin, F.; Putaux, J.-L.; Jean, B. Temperature-Controlled Star-Shaped Cellulose Nanocrystal Assemblies Resulting from Asymmetric Polymer Grafting. *ACS Macro Lett.* **2019**, *8*, 345–351. [CrossRef]
87. Musino, D.; Rivard, C.; Landrot, G.; Novales, B.; Rabilloud, T.; Capron, I. Hydroxyl groups on cellulose nanocrystal surfaces form nucleation points for silver nanoparticles of varying shapes and sizes. *J. Colloid Interface Sci.* **2021**, *584*, 360–371. [CrossRef] [PubMed]
88. Sharma, V.K.; Yngard, R.A.; Lin, Y. Silver nanoparticles: Green synthesis and their antimicrobial activities. *Adv. Colloid Interface Sci.* **2009**, *145*, 83–96. [CrossRef] [PubMed]
89. Matsuhisa, N.; Chen, X.; Bao, Z.; Someya, T. Materials and structural designs of stretchable conductors. *Chem. Soc. Rev.* **2019**, *48*, 2946–2966. [CrossRef] [PubMed]
90. Jing, X.; Li, H.; Mi, H.-Y.; Liu, Y.-J.; Feng, P.-Y.; Tan, Y.-M.; Turng, L.-S. Highly transparent, stretchable, and rapid self-healing polyvinyl alcohol/cellulose nanofibril hydrogel sensors for sensitive pressure sensing and human motion detection. *Sens. Actuators B Chem.* **2019**, *295*, 159–167. [CrossRef]
91. Chen, Y.; Zhu, J.; Yu, H.-Y.; Li, Y. Fabricating robust soft-hard network of self-healable polyvinyl alcohol composite films with functionalized cellulose nanocrystals. *Compos. Sci. Technol.* **2020**, *194*, 108165. [CrossRef]
92. Zhang, Y.; Shang, Z.; Shen, M.; Chowdhury, S.P.; Ignaszak, A.; Sun, S.; Ni, Y. Cellulose nanofibers/reduced graphene oxide/polypyrrole aerogel electrodes for high-capacitance flexible all-solid-state supercapacitors. *ACS Sustain. Chem. Eng.* **2019**, *7*, 11175–11185. [CrossRef]
93. Cao, L.; Liu, W.; Wang, L. Developing a green and edible film from Cassia gum: The effects of glycerol and sorbitol. *J. Clean. Prod.* **2018**, *175*, 276–282. [CrossRef]
94. Cao, L.; Ge, T.; Meng, F.; Xu, S.; Li, J.; Wang, L. An edible oil packaging film with improved barrier properties and heat sealability from cassia gum incorporating carboxylated cellulose nano crystal whisker. *Food Hydrocoll.* **2020**, *98*, 105251. [CrossRef]
95. Chu, Y.; Song, R.; Zhang, L.; Dai, H.; Wu, W. Water-dispersible, biocompatible and fluorescent poly (ethylene glycol)-grafted cellulose nanocrystals. *Int. J. Biol. Macromol.* **2020**, *153*, 46–54. [CrossRef]
96. Guvendiren, M.; Molde, J.; Soares, R.M.; Kohn, J. Designing biomaterials for 3D printing. *ACS Biomater. Sci. Eng.* **2016**, *2*, 1679–1693. [CrossRef] [PubMed]
97. Parak, A.; Pradeep, P.; du Toit, L.C.; Kumar, P.; Choonara, Y.E.; Pillay, V. Functionalizing bioinks for 3D bioprinting applications. *Drug Discov. Today* **2019**, *24*, 198–205. [CrossRef] [PubMed]
98. Kumar, A.; Matari, I.A.I.; Han, S.S. 3D printable carboxylated cellulose nanocrystal-reinforced hydrogel inks for tissue engineering. *Biofabrication* **2020**, *12*, 025029. [CrossRef] [PubMed]
99. Pickering, S.U. CXCVI.—Emulsions. *J. Chem. Soc. Trans.* **1907**, *91*, 2001–2021. [CrossRef]
100. Yang, Y.; Fang, Z.; Chen, X.; Zhang, W.; Xie, Y.; Chen, Y.; Liu, Z.; Yuan, W. An overview of Pickering emulsions: Solid-particle materials, classification, morphology, and applications. *Front. Pharmacol.* **2017**, *8*, 287. [CrossRef]
101. Saidane, D.; Perrin, E.; Cherhal, F.; Guellec, F.; Capron, I. Some modification of cellulose nanocrystals for functional Pickering emulsions. *Philos. Trans. R. Soc. A Math. Phys. Eng. Sci.* **2016**, *374*. [CrossRef]
102. Mikulcová, V.; Bordes, R.; Minařík, A.; Kašpárková, V. Pickering oil-in-water emulsions stabilized by carboxylated cellulose nanocrystals—Effect of the pH. *Food Hydrocoll.* **2018**, *80*, 60–67. [CrossRef]
103. Koshani, R.; van de Ven, T.G.M.; Madadlou, A. Characterization of Carboxylated Cellulose Nanocrytals Isolated through Catalyst-Assisted H2O2 Oxidation in a One-Step Procedure. *J. Agric. Food Chem.* **2018**, *66*, 7692–7700. [CrossRef]
104. Kaushik, M.; Moores, A. Nanocelluloses as versatile supports for metal nanoparticles and their applications in catalysis. *Green Chem.* **2016**, *18*, 622–637. [CrossRef]
105. Lam, E.; Hrapovic, S.; Majid, E.; Chong, J.H.; Luong, J.H. Catalysis using gold nanoparticles decorated on nanocrystalline cellulose. *Nanoscale* **2012**, *4*, 997–1002. [CrossRef]
106. Chen, L.; Cao, W.; Quinlan, P.J.; Berry, R.M.; Tam, K.C. Sustainable catalysts from gold-loaded polyamidoamine dendrimer-cellulose nanocrystals. *ACS Sustain. Chem. Eng.* **2015**, *3*, 978–985. [CrossRef]

107. Wang, Y.; He, X.; Lu, Q. Polyamidoamine dendrimer functionalized cellulose nanocrystals for CO_2 capture. *Cellulose* **2021**, *28*, 4241–4251. [CrossRef]
108. Liu, J.; Plog, A.; Groszewicz, P.; Zhao, L.; Xu, Y.; Breitzke, H.; Stark, A.; Hoffmann, R.; Gutmann, T.; Zhang, K. Design of a Heterogeneous Catalyst Based on Cellulose Nanocrystals for Cyclopropanation: Synthesis and Solid-State NMR Characterization. *Chem. A Eur. J.* **2015**, *21*, 12414–12420. [CrossRef]
109. Hu, L.; Xie, Q.; Tang, J.; Pan, C.; Yu, G.; Tam, K.C. Co (III)-Salen immobilized cellulose nanocrystals for efficient catalytic CO2 fixation into cyclic carbonates under mild conditions. *Carbohydr. Polym.* **2021**, *256*, 117558. [CrossRef]
110. Ellebracht, N.C.; Jones, C.W. Amine functionalization of cellulose nanocrystals for acid–base organocatalysis: Surface chemistry, cross-linking, and solvent effects. *Cellulose* **2018**, *25*, 6495–6512. [CrossRef]
111. Ellebracht, N.C.; Jones, C.W. Optimized cellulose nanocrystal organocatalysts outperform silica-supported analogues: Cooperativity, selectivity, and bifunctionality in acid–base aldol condensation reactions. *ACS Catal.* **2019**, *9*, 3266–3277. [CrossRef]
112. Pinto, A.H.; Taylor, J.K.; Chandradat, R.; Lam, E.; Liu, Y.; Leung, A.C.; Keating, M.; Sunasee, R. Wood-based cellulose nanocrystals as adsorbent of cationic toxic dye, Auramine O, for water treatment. *J. Environ. Chem. Eng.* **2020**, *8*, 104187. [CrossRef]
113. Samadder, R.; Akter, N.; Roy, A.C.; Uddin, M.M.; Hossen, M.J.; Azam, M.S. Magnetic nanocomposite based on polyacrylic acid and carboxylated cellulose nanocrystal for the removal of cationic dye. *RSC Adv.* **2020**, *10*, 11945–11956. [CrossRef]
114. He, X.; Male, K.B.; Nesterenko, P.N.; Brabazon, D.; Paull, B.; Luong, J.H.T. Adsorption and Desorption of Methylene Blue on Porous Carbon Monoliths and Nanocrystalline Cellulose. *ACS Appl. Mater. Interfaces* **2013**, *5*, 8796–8804. [CrossRef]
115. Batmaz, R.; Mohammed, N.; Zaman, M.; Minhas, G.; Berry, R.M.; Tam, K.C. Cellulose nanocrystals as promising adsorbents for the removal of cationic dyes. *Cellulose* **2014**, *21*, 1655–1665. [CrossRef]
116. Hang, Y.; Liu, G.; Huang, K.; Jin, W. Mechanical properties and interfacial adhesion of composite membranes probed by in-situ nano-indentation/scratch technique. *J. Membr. Sci.* **2015**, *494*, 205–215. [CrossRef]
117. Wang, J.-J.; Yang, H.-C.; Wu, M.-B.; Zhang, X.; Xu, Z.-K. Nanofiltration membranes with cellulose nanocrystals as an interlayer for unprecedented performance. *J. Mater. Chem. A* **2017**, *5*, 16289–16295. [CrossRef]
118. Liu, Y.; Bai, L.; Zhu, X.; Xu, D.; Li, G.; Liang, H.; Wiesner, M.R. The role of carboxylated cellulose nanocrystals placement in the performance of thin-film composite (TFC) membrane. *J. Membr. Sci.* **2021**, *617*, 118581. [CrossRef]
119. Yu, X.; Tong, S.; Ge, M.; Wu, L.; Zuo, J.; Cao, C.; Song, W. Adsorption of heavy metal ions from aqueous solution by carboxylated cellulose nanocrystals. *J. Environ. Sci.* **2013**, *25*, 933–943. [CrossRef]
120. Yu, H.-Y.; Zhang, D.-Z.; Lu, F.-F.; Yao, J. New Approach for Single-Step Extraction of Carboxylated Cellulose Nanocrystals for Their Use as Adsorbents and Flocculants. *ACS Sustain. Chem. Eng.* **2016**, *4*, 2632–2643. [CrossRef]
121. Guan, Y.; Yu, H.-Y.; Abdalkarim, S.Y.H.; Wang, C.; Tang, F.; Marek, J.; Chen, W.-L.; Militky, J.; Yao, J.-M. Green one-step synthesis of ZnO/cellulose nanocrystal hybrids with modulated morphologies and superfast absorption of cationic dyes. *Int. J. Biol. Macromol.* **2019**, *132*, 51–62. [CrossRef] [PubMed]
122. Awang, N.; Ramasamy, D.; Kadirgama, K.; Najafi, G.; Sidik, N.A.C. Study on friction and wear of Cellulose Nanocrystal (CNC) nanoparticle as lubricating additive in engine oil. *Int. J. Heat Mass Transf.* **2019**, *131*, 1196–1204. [CrossRef]
123. Samylingam, L.; Anamalai, K.; Kadirgama, K.; Samykano, M.; Ramasamy, D.; Noor, M.; Najafi, G.; Rahman, M.; Xian, H.W.; Sidik, N.A.C. Thermal analysis of cellulose nanocrystal-ethylene glycol nanofluid coolant. *Int. J. Heat Mass Transf.* **2018**, *127*, 173–181. [CrossRef]
124. Li, M.-C.; Ren, S.; Zhang, X.; Dong, L.; Lei, T.; Lee, S.; Wu, Q. Surface-chemistry-tuned cellulose nanocrystals in a bentonite suspension for water-based drilling fluids. *ACS Appl. Nano Mater.* **2018**, *1*, 7039–7051. [CrossRef]
125. Wu, Q.; Li, M. Modified Cellulose Nanocrystals and Their Use in Drilling Fluids. U.S. Patent US20190127625A1, 2 May 2019.
126. Li, M.-C.; Wu, Q.; Lei, T.; Mei, C.; Xu, X.; Lee, S.; Gwon, J. Thermothickening Drilling Fluids Containing Bentonite and Dual-Functionalized Cellulose Nanocrystals. *Energy Fuels* **2020**, *34*, 8206–8215. [CrossRef]
127. Sahoo, K.; Bergman, R.; Alanya-Rosenbaum, S.; Gu, H.; Liang, S. Life cycle assessment of forest-based products: A review. *Sustainability* **2019**, *11*, 4722. [CrossRef]

Review

3D-Printable Nanocellulose-Based Functional Materials: Fundamentals and Applications

Abraham Samuel Finny, Oluwatosin Popoola and Silvana Andreescu *

Department of Chemistry and Biomolecular Science, Clarkson University, Potsdam, New York, NY 13699-5810, USA; finnya@clarkson.edu (A.S.F.); popoolos@clarkson.edu (O.P.)
* Correspondence: eandrees@clarkson.edu

Abstract: Nanomaterials obtained from sustainable and natural sources have seen tremendous growth in recent times due to increasing interest in utilizing readily and widely available resources. Nanocellulose materials extracted from renewable biomasses hold great promise for increasing the sustainability of conventional materials in various applications owing to their biocompatibility, mechanical properties, ease of functionalization, and high abundance. Nanocellulose can be used to reinforce mechanical strength, impart antimicrobial activity, provide lighter, biodegradable, and more robust materials for packaging, and produce photochromic and electrochromic devices. While the fabrication and properties of nanocellulose are generally well established, their implementation in novel products and applications requires surface modification, assembly, and manufacturability to enable rapid tooling and scalable production. Additive manufacturing techniques such as 3D printing can improve functionality and enhance the ability to customize products while reducing fabrication time and wastage of materials. This review article provides an overview of nanocellulose as a sustainable material, covering the different properties, preparation methods, printability and strategies to functionalize nanocellulose into 3D-printed constructs. The applications of 3D-printed nanocellulose composites in food, environmental, and energy devices are outlined, and an overview of challenges and opportunities is provided.

Keywords: nanocellulose; 3D printing; composites; packaging; sustainable materials; additive manufacturing

Citation: Finny, A.S.; Popoola, O.; Andreescu, S. 3D-Printable Nanocellulose-Based Functional Materials: Fundamentals and Applications. *Nanomaterials* **2021**, *11*, 2358. https://doi.org/10.3390/nano11092358

Academic Editors: Rajesh Sunasee and Karina Ckless

Received: 24 August 2021
Accepted: 9 September 2021
Published: 11 September 2021

Publisher's Note: MDPI stays neutral with regard to jurisdictional claims in published maps and institutional affiliations.

1. Introduction

Nanocellulose, derived from cellulose in the form of cellulose nanocrystals, cellulose nanofibers, or bacterial nanocellulose, has emerged as a robust, new, and versatile material for energy, food, environmental, and biomedical applications [1,2]. Nanocellulose is produced from wood pulp by acid hydrolysis, and the resulting structures have dimensions ranging between 5–100 nm or longer. Its unique properties include a high surface area, high mechanical strength and barrier properties, good biocompatibility, biodegradability, and antimicrobial activity, making it a promising candidate for use in various applications. While the properties of nanocellulose are only beginning to be explored, its promise to replace traditional materials and enable the rapid manufacturing of novel products and applications is high [2]. For nanocellulose to reach its full potential as an alternative sustainable solution to traditional materials, research is needed to develop methods that can convert raw nanocellulose materials into functional constructs made of nanocellulose composites. The latest developments in manufacturing techniques such as additive manufacturing and 3D printing represent a new wave of technology that can help new developments and the realization of these composites and resulting products on an industrial scale. Figure 1 illustrates the transformation of nanocellulose from a raw material derived from plant cells into 3D-printed constructs and the different stages of the transformation process.

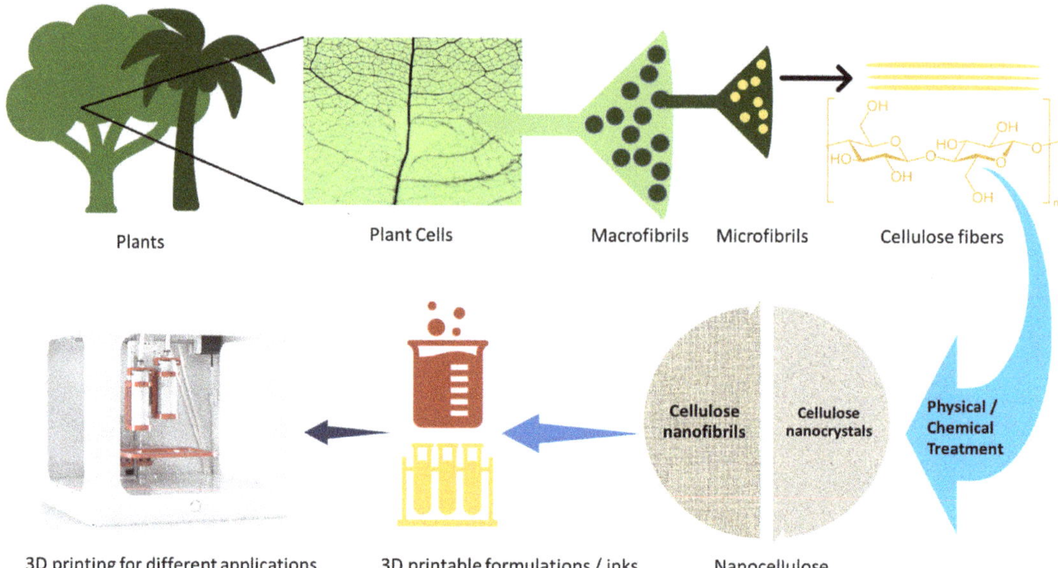

Figure 1. Schematic diagram of nanocellulose: source, processing, and 3D printability.

Cellulose, poly(1,4-*D*-glucose), is one of the most widespread natural renewable resources on Earth and represents a potential source of high-value products. Cellulose is abundant, can be produced inexpensively on a large scale from sustainable sources, is biocompatible and biodegradable, and can be modified for a variety of applications [3]. Nanocellulose is produced from cellulose by varying chemical and biological processes. Its physical properties, chemical functionality, large surface area, biocompatibility, and biodegradability make nanocellulose a promising material for a broad range of applications. Due to its versatility, nanocellulose can be used as an unlimited source for advanced functional materials, but in order to fully exploit its potential, it is critical to develop strategies to produce derivatives with improved properties in order to expand the range of potential applications. However, nanocellulosic materials do not have sufficient functional properties for advanced applications, and therefore to improve their value and utility, they require modification. Their tailorable functionality can be tuned by selecting surface modifiers or other materials such as natural or synthetic polymers to fabricate composite functional structures and useable products.

The intrinsic properties, size, and composition of native cellulose can be modified by chemical (e.g., solvent) or thermal treatment or by creating hybrid structures in conjunction with other natural or synthetic polymers and biomolecules including epoxy, starch, polyurethane, polylactic acid, hydroxyethyl cellulose, cellulose acetate butyrate, melamine-formaldehyde resins, chitosan, polystyrene, polypyrrole, methacrylates, polyacrylamide, polyacrylate, polyaniline, polyfuryl alcohol, and polyvinyl alcohol [4,5]. The properties of the resulting cellulose-based nanocomposites depend on the structural features of the starting cellulosic materials, which vary heavily on the properties of the cellulose fiber network rather than the individual nanofibers. The bonding and adhesion between the fibers are dependent on the homogeneity and origin of the nanocellulose matrix. To improve the homogeneity of nanocellulose composites, surface functionalization by introducing stable negative or positive electrostatic charges has been used to tweak the surface energy characteristics and graft functional molecules by in situ modification within the nanocellulose matrix [6]. Many hydroxyl groups available on the surface make functionalization possible while maintaining the hydrophobicity of the surface and keeping the composites thermodynamically stable [7].

There are multiple ways in which cellulosic nanocomposites can be produced. Standard methods include casting, electrospinning, melt extrusion, molding, milling, and precipitation. The recent introduction of 2D and 3D printing techniques provides an avenue for scalable and reproducible manufacturing of new materials and devices. These additive manufacturing techniques enable the production of constructs with tailorable structures and enhanced mechanical and chemical properties. Using 3D printing techniques, mechanically stable constructs are fabricated through computer-controlled deposition of materials, layer by layer, until structures with precise macroscopic and microscopic control are created [8]. Increased usage of extrusion-based 3D printing has been seen, which involves the utilization of hydrogels or printable pastes constructed by incorporating viscous materials that can retain their shape after deposition through chemical, photo, or thermal crosslinking methods [9,10]. Nanocellulose can be introduced to these pastes to facilitate the deposition and realization of stable constructs for applications. The selection of materials and their rheological properties decide whether these could be 3D printed and incorporated into applications.

The interest in nanocellulose research has grown substantially in recent years as evidenced by the exponential growth in the number of publications, with the highest numbers from countries such as China, USA, Sweden, and Canada, who have invested heavily in efforts to support reutilization and repurposing of cellulosic materials to promote the development of high-value wood-based products (Figure 2). In the past four years, with the recent growing interest in additive manufacturing, there is also increased use of 3D printing for nanocellulose research. This review article summarizes the potential of nanocellulose as a promising material for the fabrication of functional nanostructures and devices fabricated by 3D printing. Some of the commonly used bioprinters are the EnvisionTec 3D-Bioplotter, RegenHu 3DDiscovery Evolution, Poietis NGB-R and NGB-C 3D bioprinters, Cellink Bio X6, 3D Systems Allevi Series, Rokit Dr. Invivo 4D, Inventia Rastrum, Organovo NovoGen MMX, Aspect Biosystems RX1, and Bioprinting Solutions Fabion 2. Even though most of these printers are used for printing functional tissues and organs, in recent times they have seen a tremendous increase for use in other areas for creating functional materials (Figure 2). The properties, modification, and fabrication of nanocellulose-based nanocomposites and printable inks for food, energy, and environmental applications have seen increased interest in recent times.

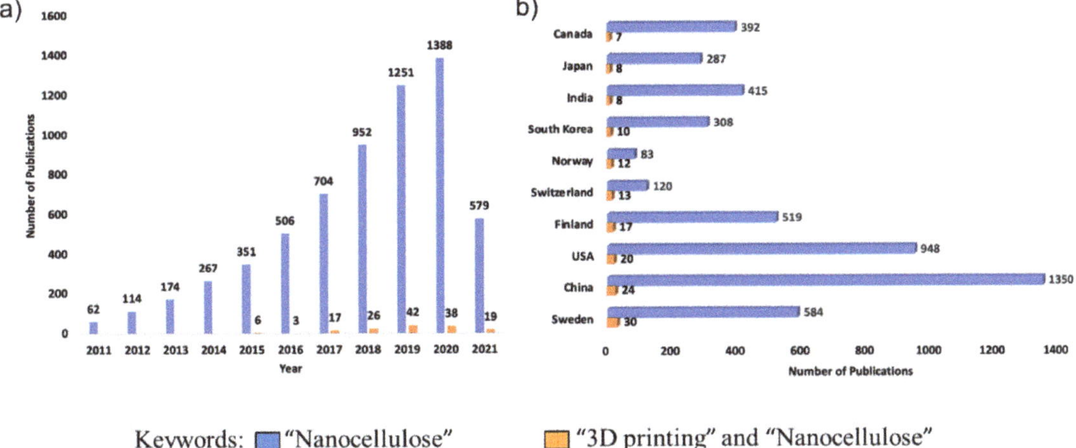

Figure 2. Graphs showing the number of publications between 2011–2021 (**a**) and publications per country (**b**) using Web of Science (accessed on 30 June 2021).

2. Nanocellulose: Preparation, Treatment, Functionality, and 3D Printability

Nanocellulosic materials include three types of structures [7]: (i) bacterial nanocellulose (BNC), bacterial cellulose (BC), or microbial cellulose; (ii) nanofibrillated cellulose (NFC), also called cellulose nanofiber (CNF) or cellulose nanofibrils; and (iii) nanocrystalline cellulose (NCC), also known as crystallites, whiskers, or rod-like cellulose microcrystals as well as crystalline nanocellulose [11–13]. Nanocellulose is considered non-toxic, renewable, biodegradable, and safe [1,14]. Its high aspect ratio gives good strength and toughness as a reinforcement matrix for various formulations in additive manufacturing, food packaging, medical, flexible electronics, military, and energy applications [15–19]. This section discusses the main functional properties of nanocellulose, modification techniques, and composites to create 3D-printable structures.

2.1. Preparation

Cellulose is reported to be the most abundant polysaccharide in nature [20]. It is widely distributed as it constitutes the major component of a plant's cell wall; hence, cellulose can be easily sourced from plants, bacterial pellicles, and agricultural wastes [21,22]. The diameter of BC ranges between 20–100 nm, and it is commonly produced from low-molecular-weight sugars and alcohols. Glucose chains are formed inside the bacterial body during the biosynthesis process and extrude out through tiny pores present on the cell envelope [23]. These chains combine to form microfibrils further aggregated into ribbons to generate a web-shaped network structure with BC [23]. Stefan-Ovidiu D. et al. also reported BNC production from a kombucha membrane obtained as a secondary product from the fermentation of tea broth in beverage production [24]. Several processing methods have been used to produce CNC and CNF from cellulose sources, usually agricultural wastes, by exposing the basic building blocks of cellulose [20]. These methods can be summarized into two: (i) pretreatment or purification of the agricultural biomass for fractionation and (ii) treatment with the isolation of nanocellulose from purified cellulose [25]. The plant cell wall contains lignin and hemicellulose, which serve as a protective coating around the cellulose fibers, and this enables the cell wall to maintain its structured form. To extract the cellulose to produce nanocellulose, a hybrid preparation system is usually employed to remove the protective coating through mechanical forces followed by chemical pretreatment [22]. The chemical pretreatment step involves using chemicals such as alkali, the most common method in the industrial production of cellulose, which enables the effective delignification and partial solvation of the hemicellulose. Common alkalis used for this purpose include NaOH, KOH, and NH_4OH [26,27].

Other chemical pretreatment methods include oxidative delignification, which involves converting lignin copolymer into carboxylic acids using strong oxidizers such as hydrogen peroxide or ozone. To further ease the penetration of hydrolytic media, oxidative delignification could be followed by effective enzymatic hydrolysis. Due to the low cost of hydrogen peroxide and its effectiveness in delignification pretreatment, the method is widely used on woody biomass. Several combined pretreatment steps are often needed to achieve nanosized materials with a high aspect ratio and crystallinity index [26,27]. Khan et al. produced CNC with an aspect ratio of 10.45 ± 3.44 nm and a crystallinity index of 66.7% by pretreating dunchi fiber with alkali, hydrogen peroxide, and sulfuric acid sequentially [28].

Other works have highlighted the use of ionic liquids (ILs) as a selective pretreatment method for removing the lignin and hemicellulose (soluble) from lignocellulose biomass. The method enables precipitation of the insoluble cellulose, which can then be filtered off and washed. ILs are expensive and have been found to inactivate cellulase enzymes irreversibly; hence it may not be suitable for pretreating cellulose for biological uses [29]. Deep eutectic solvents (DESs) are recently emerging solvents with promising prospects in lignocellulose biomass pretreatment. DESs have similar physicochemical characteristics as ILs and can be prepared at a lesser cost. While ILs are produced from the combinations of particular types of discrete anions and cations, DESs are formed by mixing Bronsted or Lewis acid and bases made up of diverse cationic and anionic species [30], and their physicochemical properties can be tuned to a specific need by adjusting the molar ratio of each constituent [31].

Following the isolation of the nanosized cellulose particles from the pretreated fiber, a sequential chemical, enzymatic, or mechanical treatment is applied. The selection of suitable nanocellulose isolation methods depends primarily on the desired degree of crystallinity and the type of nanocellulose (CNF or CNC) of interest [25]. Fan B. et al. reported the preparation of nanocellulose from the bamboo leaf that was initially pretreated by grinding, sieving, dispersion in water, ultrasonication, oven-drying, functionalization, and then subjected to further mechanical ultrasonication treatment [32]. Fortunati, E. et al. obtained nanofiber with about 10 nm diameter and 175 nm length from a sunflower using sulfuric acid treatment [33]. CNF of ~100 nm and 67% crystallinity was derived through acid hydrolysis of pretreated ginger tuber by Abral H. et al.; however, the crystallinity dropped to 48% after ultrasonication. A common technique is the direct acid hydrolysis of microcrystalline cellulose (MCC), microfibrillated cellulose (MFC), nanofibrillated cellulose (NFC), wood fiber (WF), or plant fiber (PF) to produce highly crystalline needle-like particles with a high aspect ratio of about 10 nm width and several 100 nm in length and made of about 100% cellulose with a high fraction of Iβ crystal structure [1]. A pure form of BC can be synthesized using bacteria (such as *Acetobacter xylinum*) without pretreatment to remove unwanted contaminants or impurities such as hemicellulose, pectin, and lignin [34].

2.2. Functional Nanocellulose-Based Composite Nanostructures

Nanostructured materials with defined structures (size, shape, and connectivity) and controllable physicochemical properties (sorption and separation) are of great interest in materials science, opening new avenues for designing functional materials and devices [34–38]. Their advantages include a high surface-to-volume ratio, large pore volume, and the possibility to control charge, accessibility, and availability of surface functional groups. Changes in these properties, i.e., volume, porosity, and shape, can be triggered by varying environmental conditions, pH, ionic strength, water, electricity, or light [39]. This has allowed the development of stimuli-responsive materials with reversible and programmable actuation and can offer promising capabilities in sensing, artificial muscles, and electronic devices [40].

The combination of nanocellulose with other materials, polymers, and fillers can increase mechanical and thermal stability and change the wetting behavior of the native material [41]. For example, significant enhancement of cellulose tensile strength was achieved by loading cellulose with only 0.2% graphene [42]. The ability of nanocellulose to enhance mechanical properties has enabled their use for the fabrication of printable electronics [43] or sustainable food packaging, as a new type of packaging material [44]. Other possibilities include using polymers to change the surface properties, porosity, and wetting characteristics. Water permeation and salt rejection capabilities have been achieved by interfacial polymerization of amino-functional piperazine and 1,3,5-trimesolyl chloride on cellulose, which enabled applications in water purification [45]. Ultrathin nanocellulose shell microparticles were obtained via emulsion-templated colloidal assembly. This method enabled the loading of pH-responsive compounds, such as methylene blue, using polystyrene as a matrix. The carboxyl groups of the CNF shell showed pH-dependent properties and strong adsorption and drug release behavior [46]. Chemical or biological receptors can also be integrated to provide selectivity for sensing applications. In this case, fabrication requires appropriate surface chemistry suitable for the assembly of biomolecules to maintain bioactivity for selective binding.

2.3. Three-Dimensional Printability of Nanocellulose Composites

Three-dimensional printing is an emerging technology that applies rapid prototyping techniques to fabricate computer-designed models and integrate complex structures with an unprecedented level of functionality for applications. The method is currently being explored for various materials and devices, and has proven helpful for patterning and assembling nanomaterials in geometrically complex functional structures [47]. The method provides unique opportunities for the large-scale manufacturing of nanocellulose-based constructs and nanocomposite structures. However, achieving the required level of functionality and integration is challenging as it requires the ability to pattern and develop nanocellulose-based inks of suitable rheology, composition, deposition, and crosslinking conditions for realizing printability. Unlike conventional plastics, nanocellulose is typically processed in aqueous solutions, which are used to create printable inks but might result in anisotropic deformation and poor mechanical properties [48]. Therefore, achieving high consistency and ensuring printability with shape retention and structure by 3D printing is challenging. Nevertheless, the integration of nanocellulose with 3D printing can enable the creation of functional structures with optical, mechanical, electrical, and biological properties for many applications. In contrast with conventional deposition methods, 3D printing enables rapid prototyping and provides mass production capabilities, reducing development time, fabrication steps, and cost.

Printing of 3D nanocellulose-based devices requires incorporating additional materials, typically polymers and solvents, in conjunction with nanocellulose. The printing process also involves post-processing to consolidate the 3D structure and provide shape consistency. In fabricating composite structures, the size, shape, and type of nanocellulose materials as well as the type and properties of the additional binders and their complementarity play a critical role. It has been shown that when a neat CNF ink is 3D printed into a specific shape, the printed structure collapses when handled or exposed to mechanical force due to the inability of the fibers to crosslink; hence, neat CNF ink is not conducive for 3D printing [49]. Crosslinking enables the 3D-printed material to maintain its shape and mechanical structure. To achieve improved surface properties that enhance structural sustenance, supportive chemical species can be grafted to the reactive end (-OH side groups) of the nanocellulose [1]. Several components have been combined with CNF at varying concentrations, such as biodegradable polymers and hydrogels. Reagents such as xylan-tyramine (XT) [38], which alone is not printable, can be used in a composite structure with CNF at the appropriate concentration to impact crosslinking ability (Figure 3). Silver nanoparticles (AgNPs), a broad-spectrum antimicrobial agent, have been widely used to impact the antimicrobial properties of nanocellulose and fabricate antimicrobial constructs by 3D printing [50–52].

In a recent example, the manufacturing of stable structures with cellulose nanocrystals (CNCs) involved: (1) obtaining homogenous dispersion of CNC hydrogel ink by mixing, (2) printing cellulose scaffolds by aligning the anisotropic CNC upon printing, and (3) UV curing of the printed scaffold (Figure 4) [53]. CNC was used in this example as a reinforcing agent (up to 35%), while CNF was added at a lower concentration (1%) to enhance shape retention. N-isopropyl acrylamide (NIPAM), a photo-crosslinkable polymer, was used to produce biocompatible hydrogel inks. The addition of ε-polylysine imparted antimicrobial properties for potential biomedical uses. This composition enabled the ability to control the composite structures' local orientation, mechanical properties, actuation, and printability while providing additional functionality, e.g., antimicrobial activity and the ability to bend and twist upon hydration. The nanocellulose–NIPAM hydrogels were printed using a direct ink writing procedure with the hydrogels placed in plastic cartridges and extruded through steel nozzles under compressed air. Curing was achieved under UV exposure conditions. It was observed that nanocellulose was able to constrain the swelling and shrinkage of the PNIPAM structures, enabling the production of shape-morphing nanocellulose-based composites through direct ink writing [53].

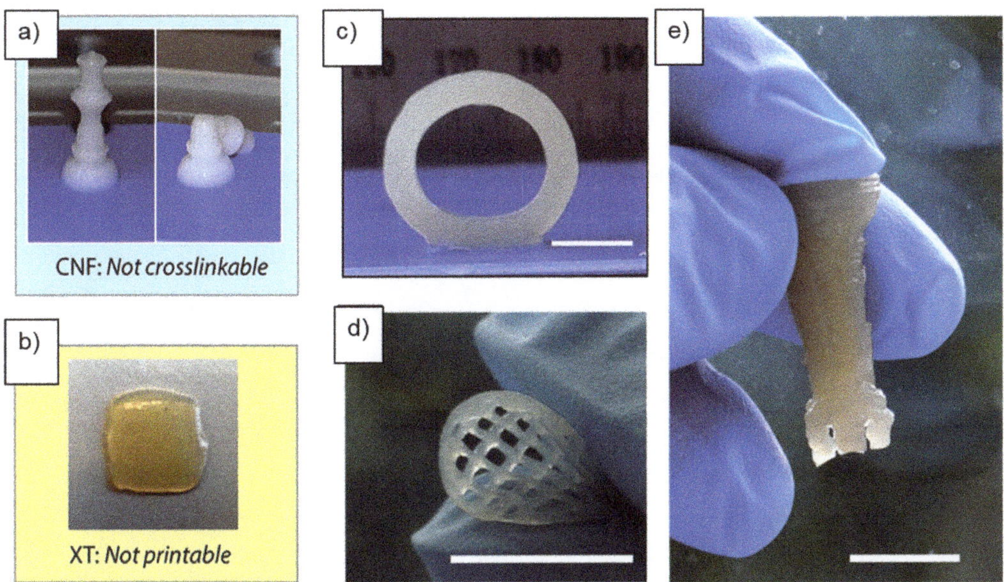

Figure 3. A queen chess piece printed with pure CNFs which collapsed due to lack of crosslinking (**a**), poor printability shown by unsuccessful grid printed with pure XT (**b**). Three-dimensional printed and crosslinked, freestanding and crosslinked cylinder (**c**) grid, handled and bent in air (**d**). Rook chess piece, held upside down from optimized formulation of CNF and XT (**e**), reprinted with permission from ref. [49]. Copyright 2017, American Chemical Society.

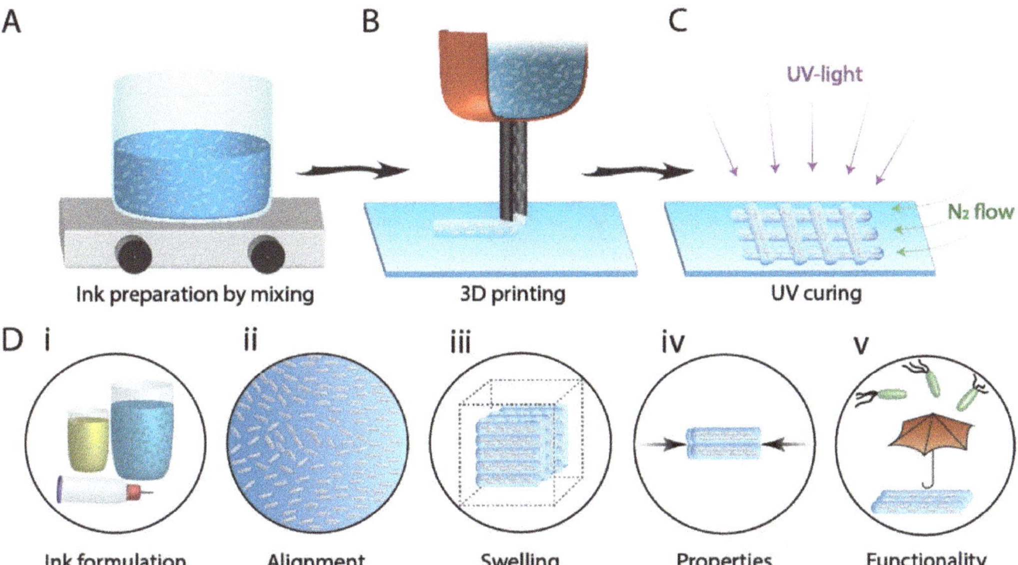

Figure 4. Illustration of the steps involved in the preparation of 3D-printed functional CNC-based hydrogels, involving: (**A**) ink formulation, (**B**) ink writing of cellulose-based polymer ink, (**C**) post-treatment to cure the printed construct, and (**D**) (**i–v**) characteristics of inks and printed constructs, reprinted from ref. [53].

In another report, a 3D-printed biobased composite made of acetylated CNCs (5–20 wt%) and poly(3-hydroxybutyrate-co-3hydroxyhexanoate) (PHBH), a biodegradable aliphatic polyester was fabricated as an environmentally friendly alternative bioplastic [54]. The method involved the use of a melt-compounding process, preceded by solvent mixing to obtained biobased composites with good dispersion behavior of PHBH and CNCs. An extrusion-based printing, fused deposition modeling (FDM) was further used to print the composite. Acetylation of the CNC was first required to create compatibility between the CNCs and the PHBH and enable the formation of a stable structure (Figure 5).

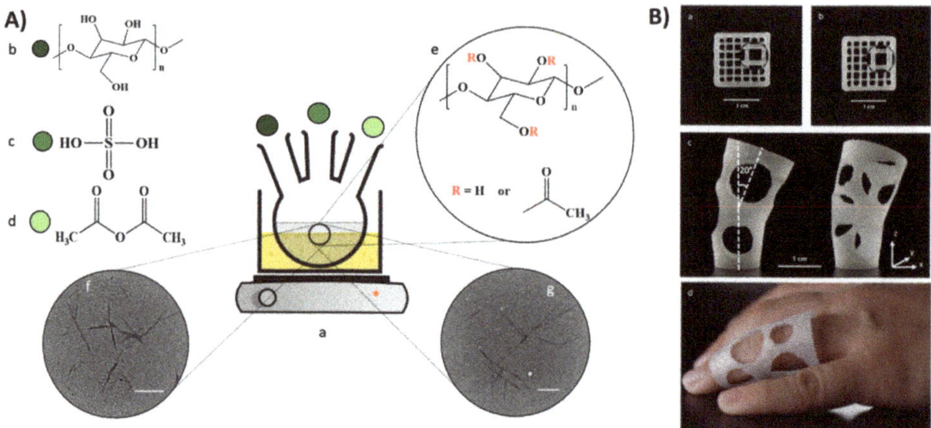

Figure 5. Illustration of CNC acetylation process and 3D-printed nanocellulose–PHBH-based composites showing the (**A**) acetylation process (**a**) of a CNC particle of a cellulose structure (**b**) by sulphuric acid (**c**) and acetic anhydride (**d**), and acetylated nanocellulose after functionalization (**e–g**). (**B**) FDM 3D-printed nanocellulose composites of PHBH-acetylated CNC (10%) showing an alteration of 0–90° (**a–c**) as an example of a medical device for finger dislocation (**c**,**d**), used with permission from ref. [54].

3. Applications of 3D-Printed Nanocellulose-Based Materials

Even though 3D-printable nanocellulose-based composites are still in their infancy, there has been an increase in their applications in different fields ranging from biomedicine, including wound dressing, drug release, and tissue engineering, sensors, food, and packaging, to energy storage and electronics, with growing interest in other areas as well [17–55], summarized in Figure 6. This section discusses the recent development in 3D-printed nanocellulose-based composites for food, environmental, food packaging, energy, and electrochemical applications.

3.1. Environmental Applications

Due to its sustainability, biocompatibility, and ability to form high mechanically robust and stable structures with a broad range of functionalities, nanocellulose has found various applications in the environmental field. Nanocellulose-based printed devices can be used as filters and membranes for water purification and environmental remediation, e.g., for pollutant removal, filtration, and desalination [18], or as materials to create sensors for environmental detection. A printed sensor on a bagasse-derived CNF-based biocomposite was demonstrated to be applicable for a wide range of humidity sensing (20–90% relative humidity, RH) with potential for soil and agricultural products monitoring in the smart farming sector [56]. Bagasse, a dry pulpy residue from the sugarcane industry, was oxidized using a 2,2,6,6-Tetramethylpiperidinyl-1-oxyl (TEMPO)-mediated oxidation process after two different pretreatment methods for comparison, (i) soda and (ii) hot water and soda, to obtain CNF. Four CNF-based sensors were then fabricated by screen printing carbon-based interdigitated electrodes (IDEs), which are reliably sensitive to impedance changes. These four sensors showed decreasing impedance with increasing RH in a humidity chamber experiment with no significant difference in sensing behavior between the four sensors. Polyethylene glycol (PEG) has been reported to enhance the mechanical properties of CNF films and the printability of CNF substrates [57,58]. To enhance the ductility of the CNF film, PEG was used to fabricate self-standing humidity sensors. The response of the sensors containing PEG was fast and stable at each level within the tested range of 7×10^5 Ω (at 20% RH) to 1.5×10^4 Ω (at 90% RH), while the counterpart without PEG, FS_T3.8, covered a more comprehensive range than was represented by the change of impedance level from 8.6×10^6 Ω (at 20% RH) to 8.3×10^3 Ω (at 90% RH), and the average values of the impedances for each RH level were used to construct the calibration curve (Figure 7). However, the incorporation of the PEG plasticizer into the biocomposite film decreased the oxygen barrier significantly.

Figure 6. Summary highlighting the broad examples of applications of 3D-printed nanocellulose constructs.

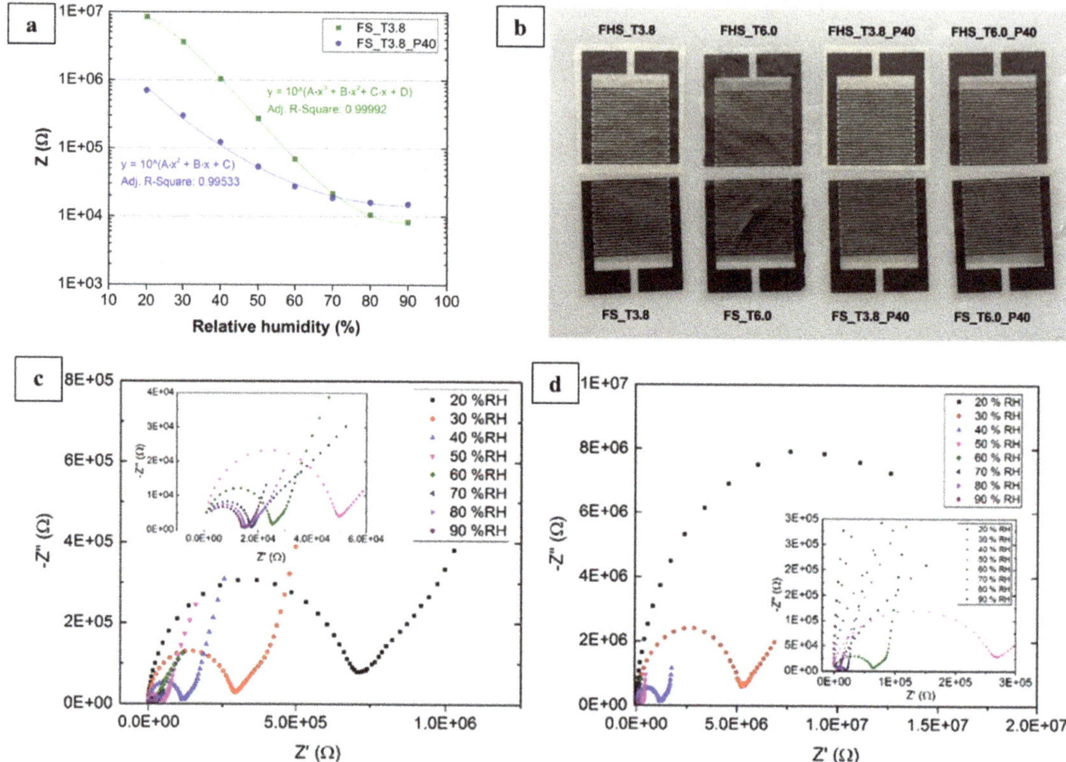

Figure 7. Calibration curves of tested samples (green, without plasticizer, and blue, with plasticizer) (**a**); RH sensors based on CNF films with screen-printed carbon electrodes (**b**). Impedance spectra under different humidity levels for the sample with plasticizer (**c**) and without plasticizer (**d**) at 25 °C. Insets: spectra for humidity levels 50, 60, 70, 80, and 90%, used with permission from ref. [53].

Another 3D-printable nanocellulose-enhanced composite of PLA/CNF (polylactic acid/cellulose nanofibers) was successfully printed (Figure 8) by T. Ambone et al., with high tensile strength, stress, and strain. This structure has potential as an eco-friendly packaging material, being fabricated from bioderived and biodegradable matrixes and fillers [59]. PLA/CNF composites were formulated by incorporating 1/3/5 wt% of CNF into the matrix of PLA to improve the poor mechanical properties of PLA and were printed through fused filament fabrication (FFF)-3D printing process. The micrographs of the printed materials from scanning electron microscopy reveal that 3% CNF in PLA was evenly distributed because no CNF aggregates were observed in the fracture surfaces. Furthermore, as indicated with black arrows, the voids, welding points, and overlapped structure (Figure 9) that were responsible for the porosity and lower mechanical properties of the fracture surface morphology of the 3D-printed neat PLA are absent in the fracture surface of PLA/3% CNF nanocomposite, while micro aggregates were seen on PLA/5%CNF. Furthermore, through differential scanning calorimetry, it was discovered that the incorporation of CNF accelerates the nucleation and enhances the crystallinity of the 3D-printed PLA. Further characterization of the PLA/CNF FFF-3D-printed material using X-ray microtomography revealed that the incorporation of the nanocellulose did not compromise the construct's thermal stability, but it did have lesser voids compared to the ordinary 3D-printed PLA.

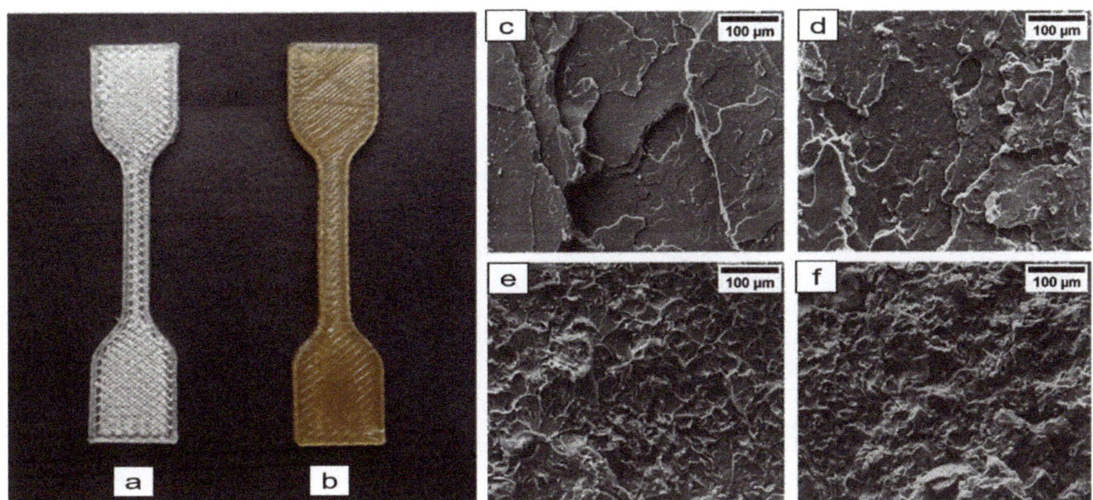

Figure 8. Three-dimensional-printed tensile specimen of PLA (**a**) and PLA/1CNF (**b**) composites, scanning electron microscopy micrographs (×600 magnification) of the tensile fractured surface of compression-molded PLA (**c**), PLA/1% CNF (**d**), PLA/3CNF (**e**), and PLA/5CNF (**f**) composites, used with permission from ref. [59].

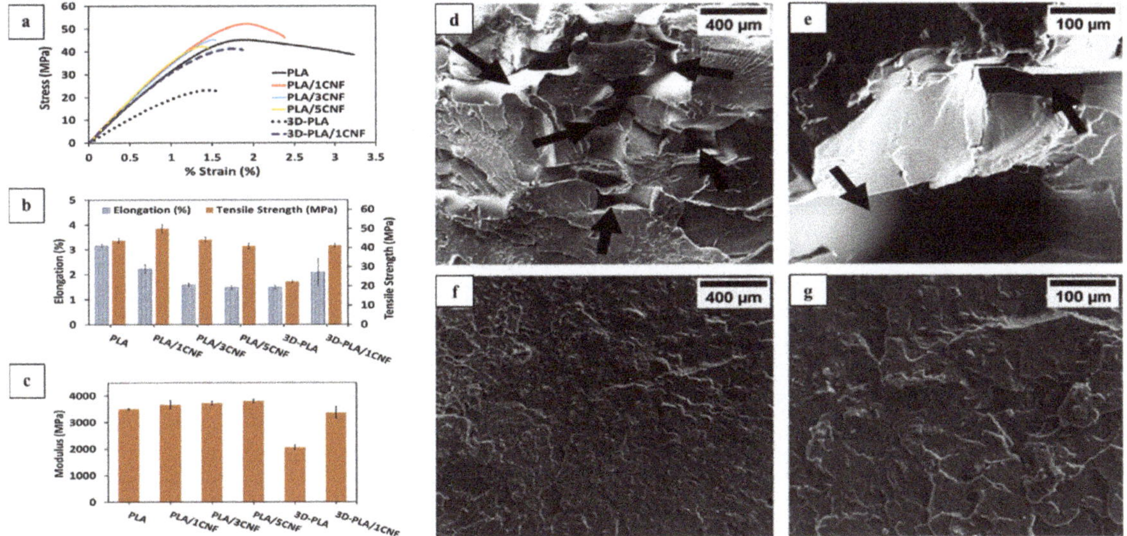

Figure 9. Tensile stress–strain curve of compression-molded and 3D-printed PLA and PLA/CNF composites (**a**) and histograms of mechanical properties of the samples (**b**,**c**). Scanning electron microscopy micrographs (×150 and ×600) of the tensile fracture surface of 3D-printed PLA (**d**,**e**) and 3D-PLA/1% cellulose nanofiber composites (**f**,**g**), used with permission from ref. [59].

Another potential growth area in the environmental field is to use 3D printing to fabricate nanocellulose-based filters and membranes for environmental remediation and cleanup. Due to its high adsorption capacity, functionality, and high specific surface area, nanocellulose is a promising sustainable adsorbent for removing chemical contaminants such as heavy metals and organic dyes. The abundance of the functional group enables further enhancement of the adsorption efficiency, making nanocellulose a particularly suitable sorbent for environmental decontamination and a possible substitute to the currently used adsorbents such as activated carbon, inorganic zeolites, silica, or polymeric-based materials [60]. Functionalization of nanocellulose can be done depending on the type of contaminant. Typical procedures include adding carboxyl-, amino-, or thiourea-using chemical modifiers such as ethylenediaminetetraacetic acid (EDTA) or carboxymethyl groups to enhance heavy metal adsorption [61]. While applications of nanocellulose for heavy metal ion and other contaminants adsorption have been reported, the use of 3D printing to create systems such as filters and membranes for water purification remains largely unexplored.

3.2. Food and Packaging Applications

With the need to replace petroleum-based products, there is a demand for sustainable, biodegradable materials and packaging to increase sustainability throughout the food chain, and many exciting materials can be developed using nanotechnology products [62]. The biocompatibility and suitable shear-thinning property of nanocellulose, even when used at a low concentration, provide unique features for this material to be used for food production, packaging, and sensing applications. Nanocellulose can reinforce and stabilize printable composites, enhance the printability of nutritious foods, and improve barrier properties when used with biopolymers or synthetic polymers [63,64]. A few selected applications featuring 3D printing techniques of nanocellulose composite materials are described in this section.

Three-dimensional-printable food using CNF-based nutritional pastes, containing ingredients from all main groups of nutrients such as carbohydrates, protein, fat, and dietary fiber was reported by Martina and co-workers [65]; these pastes were formulated from mixtures of varying percentages of CNF, starch, rye bran, oat protein concentrates (OPC), fava bean protein concentrates (FBPC), skimmed milk powder (SMP), and semi-skimmed milk powder (SSMP). A simplified extrusion pump system (Figure 10) controlled by air pressure was employed in the 3D food printing through the tip of a syringe filled with the prepared and well-mixed pasted. There was a noticeable difference in the effects of the two drying methods (oven-drying and freeze-drying) used in this research on the printed samples (Figure 10), the freeze-dried samples possess higher dry matter content but less hardness compared to oven-dried samples, which makes the freeze-drying method preferable, especially for samples with initial dry matter content less than 35%. It was observed that in samples where part of the SSMP constituent was replaced with CNF, there was a reduction in the drying rate which was attributed to the high water-binding capacity of CNF and the lower dry matter content of CNF-containing samples leading to softening of such samples. The construct printed from 0.8% CNF-50% SSMP composite was achieved without clogging of printer tips, providing high printing quality and good structural stability after freeze-drying and oven-drying. High-quality printing was also achieved with pastes containing 60% SSMP due to higher yield stress and good post-printing shape stability. The study highlighted the importance of optimizing critical parameters, such as the solid content, drying rate, and hardness, that affect the composition and the shape stability of the printed paste.

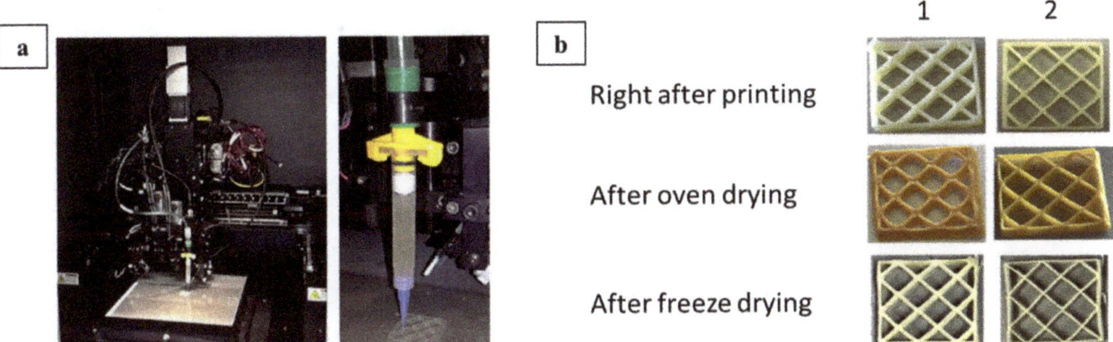

Figure 10. The material extrusion type of device used for 3D printing of food materials (**a**). The effect of the oven freeze-drying on the appearance of 3D-printed samples (**b**): 0.8% CNF + 50% SSMP (1) and 60% SSMP (2), used with permission from ref. [65].

Alternative packaging materials are also used to preserve the quality of packaged food. As much as 30% of the loss rate of fruits was attributed to mechanical damage during transportation [66]. The irreversible loss in the quality of vegetables and fruits is due to the vibration, collision, shock, and static pressure they are subjected to during their transportation [67,68]. The difficulties in degrading and recycling the commonly used foamed polyethylene (EPE) and polystyrene (EPS) packaging materials [6] call for the development of an eco-friendly biodegradable packaging material. Spoilage of fruits and vegetables due to the actions of microbes during transportation and storage is another pressing issue in the food packaging industry. A 3D-printed food packaging material with cushioning and antimicrobial dual-function from ink based on carboxymethyl nanocellulose (CMC) was recently reported [50]. First, the fibers of CMC were discovered to form entangled 3D network structures after freeze-drying, suggesting its suitability for fabricating cushioning aerogel for packaging. To prepare the CMC-based 3D-printable ink, shell ink, Irgacure 2959, and N, N′-methylenebis(acrylamide) were mixed with 8 mL of varying concentrations (75, 50, 25, and 0%) of glycerin solution, and the mixture was heated in a water bath to aid complete dissolution. The obtained solutions were added to a mixture of CMC, Irgacure 2959, and N, N′-methylenebis(acrylamide) under constant stirring. The rheology of CMC and CMC-based inks reveals the positive impact of increasing glycerin concentration. The best printability was achieved with the composite containing the higher amount of glycerin, which had the largest yield stress and highest viscosity at the lowest shear rates (Figure 11), indicating better printability. The formulated inks were printed via coaxial 3D printing technology, cured by UV for 7 min and freeze-dried. The printed sample with 75% glycerin ($CNGA_{75\text{-}7}$) fully retained its shape after freeze-drying, while the samples with lower glycerin concentrations had some degree of shrinkage and deformations. To study the effect of curing duration on crosslinking efficiency, $CNGA_{75}$ samples were exposed to UV rays for 5, 7, and 9 min and then lyophilized (Figure 12). In general, increase in stress under the same strain and the elastic modulus of the aerogels were observed as the crosslinking time increases, while their resilience and cushioning performance were decreasing. The $CNGA_{75\text{-}7}$ with 7 min curing time was suggested as a good candidate for fruit packaging. Chitosan/AgNPs inside the aerogels by coaxial 3D printing technology were used to form a core–shell fiber with a translucent matrix shell and a yellow chitosan/AgNPs core. The degradation, swelling properties, cushioning performance, and silver release behavior was evaluated. The 3D-printed cushioning packaging exhibited antibacterial potency against *Escherichia coli* and *Staphylococcus aureus* [50].

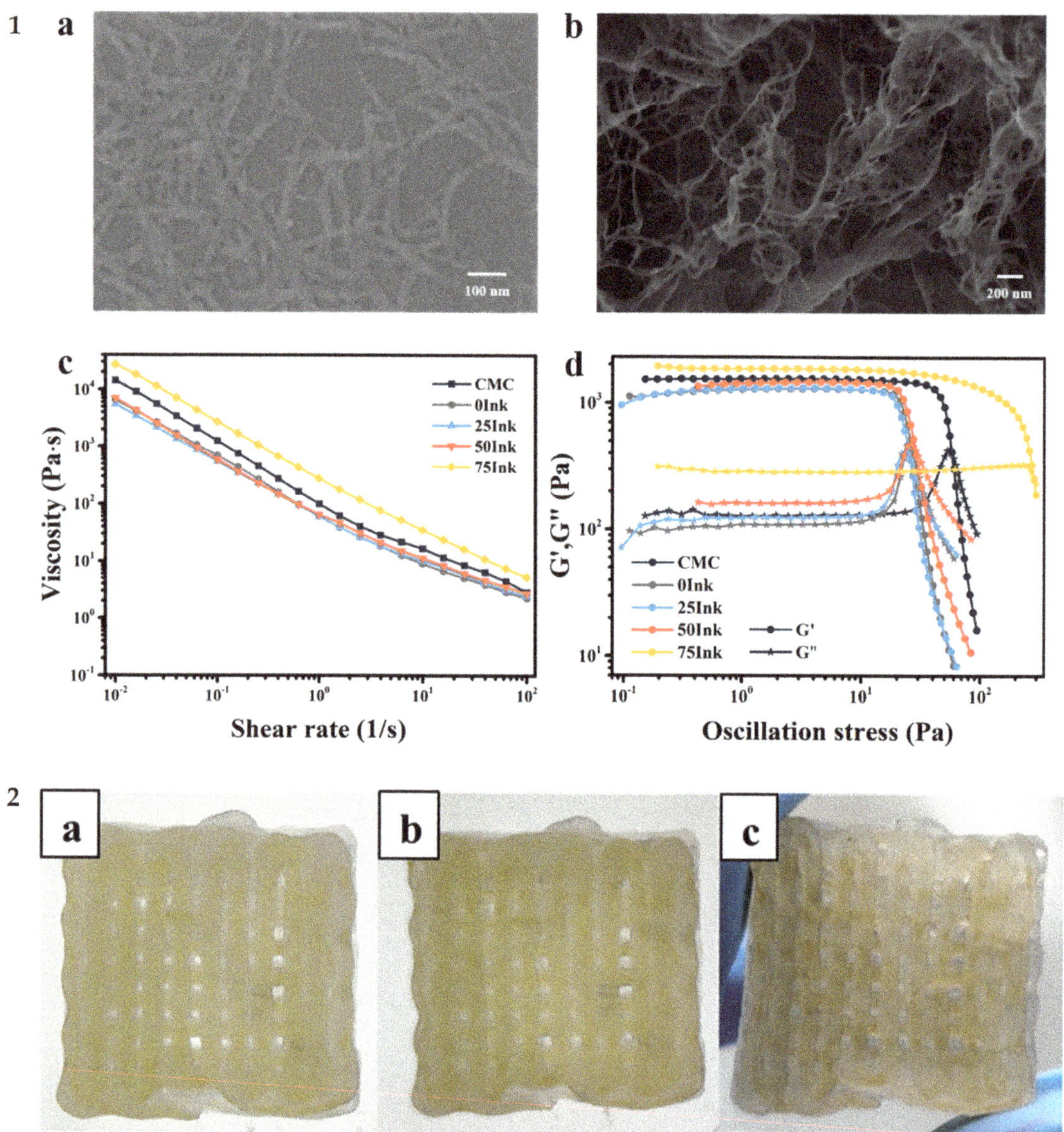

Figure 11. SEM images of CMC before (**1a**) and after freeze-drying (**1b**). Viscosity as a function of the shear rate for CMC and CMC-based inks (**1c**), G′ and G″ as a function of the shear stress for CMC and CMC-based inks (**1d**); Photographs of CMC-based 3D-printed aerogel before curing (**2a**), after curing (**2b**) and after freeze-drying (**2c**), used with permission from Ref. [50].

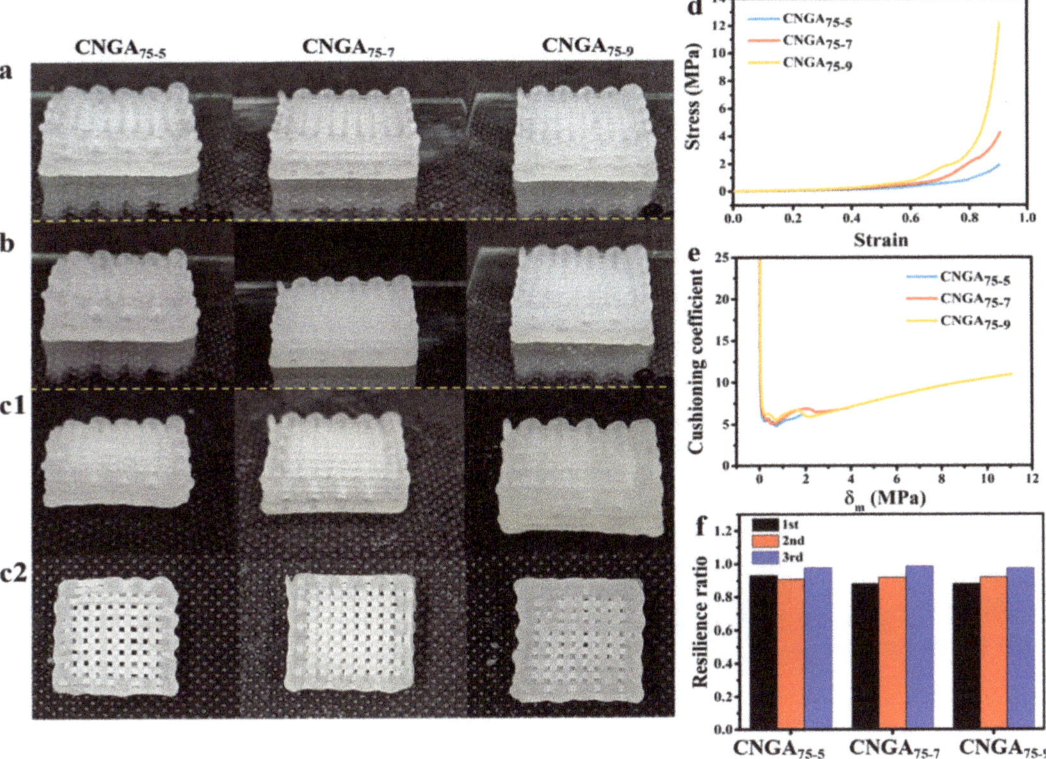

Figure 12. Morphology of 3D-printed samples by 75ink with different crosslinking times, before (**a**) and after (**b**) UV curing and after freeze-drying (**c1**,**c2**). (**d**) Stress-strain curves, (**e**) cushioning curves, and (**f**) thrice compression resilience ratio of samples with different crosslinking times, used with permission from ref. [50].

In addition to food packaging, nanocellulose hydrogels could be used to create smart indicators for intelligent packaging [62]. For example, a sugarcane-based nanocellulose hydrogel was developed in the form of a pH-responsive freshness indicator that changes color when meat deteriorates. The concept is illustrated in Figure 13, showing the formation of the Zn^{2+}-nanocellulose network prepared by induced gelation of carboxylated CNFs from cellulose filaments [44]. The nanocellulose hydrogel was used as a carrier for a pH-responsive dye and as a sorbent for CO_2 to improve the color sensitivity of the hydrogel. The indicator color was correlated with the CO_2 and microbial growth and changes in volatile compounds as a freshness indicator in packaged chicken breast. Such smart labels made of hydrogels can be easily fabricated by 3D printing using 3D extrusion bioprinting [9], which enables printing of viscous inks of viscosity up to 6 × 107 mPa.s. The viscosity and ink composition can be easily tuned to achieve printability and mechanical properties by varying the hydrogel components' composition, gelation conditions, and rheology characteristics. Using 3D printing, such sensors can be manufactured inexpensively and in large quantities.

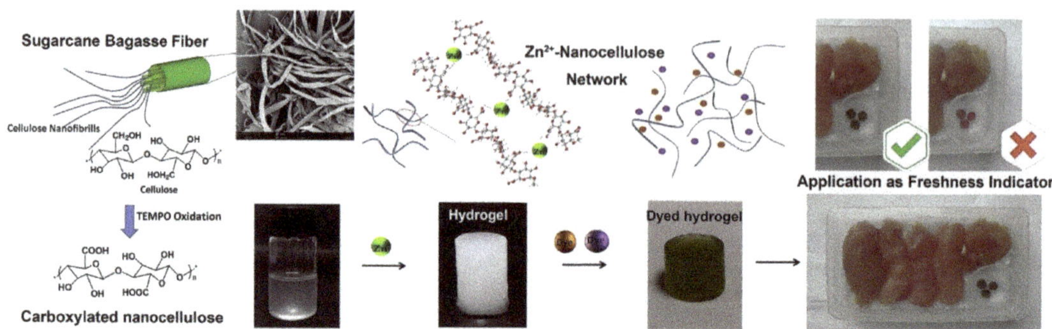

Figure 13. Design of CNF-based smart freshness indicator prepared by a 2,2,6,6-tetramethyl piperidinyl-1-oxyl (TEMPO) catalysis method for monitoring degradation in packaged meat, used with permission from ref. [44].

3.3. Energy and Electrochemical Devices

Additional applications of 3D-printed nanocellulose composites have been explored in the development of energy and electrochemical devices. In the conventional approaches, conductive inks contain expensive noble metals at a concentration of about 60% to augment the conductivity of the inks, but this high concentration of metals makes the ink toxic [69] and reactive to air and other environmental factors, which reduce its chemical stability. Nanocellulose was explored as a biocompatible material to add to the ink to improve the properties of conductive inks in an effort to reduce toxicity and improve the sustainability of energy and electrochemical devices.

Films fabricated from graphene- and nanocellulose-based inks and pastes with high chemical stability and no record of cytotoxicity were developed as metal-free conductive substitutes from water-based inks with potential for applications in the production of liquid-phase electronic device constructs [70]. These films presented interesting electronic properties, suggesting potential applications as a substitute for the metal-containing conductive inks for electric or optoelectronic devices such as sensors, supercapacitors, and solar cells. Nanocrystalline cellulose (NCC), NCC type I, and NCC type II were prepared with different stirring and heating temperature conditions using an established procedure [71]. These NCCs were used as a dispersing agent for graphene oxide (GO). Varying concentrations of carbon nanotubes (CNTs) in water were tested, and aqueous ammonia (NH_4OH) enabled a reaction medium with a basic pH suitable for the chemical crosslinking and aggregation of the graphenic nanomaterials during the hydrothermal treatment at 180 °C. The procedure provided low-viscosity inks, high-viscosity paste, or self-standing hydrogel. Conductive films were then formed by spray-coating the inks on glass substrates while the pastes were deposited via a rod-coating method on an agate rod or glass substrates. The resulting films were characterized in terms of resistivity, morphology, microstructure, thickness, and surface roughness, and it was observed that films from low-viscosity inks possessed smoother surfaces and negligibly smaller pores than the rough-surfaced films from the pastes having more prominent pores [71]. The morphology of the resulting structures, shown in Figure 14, indicates that hydrothermal treatment over 30 min always results in hydrogel formation, irrespective of the other parameters. Subjecting this hydrogel to unidirectional freezing, followed by lyophilization, leads to the formation of anisotropic porous-microstructured aerogels (Figure 15) applicable to energy and environmental remediation applications.

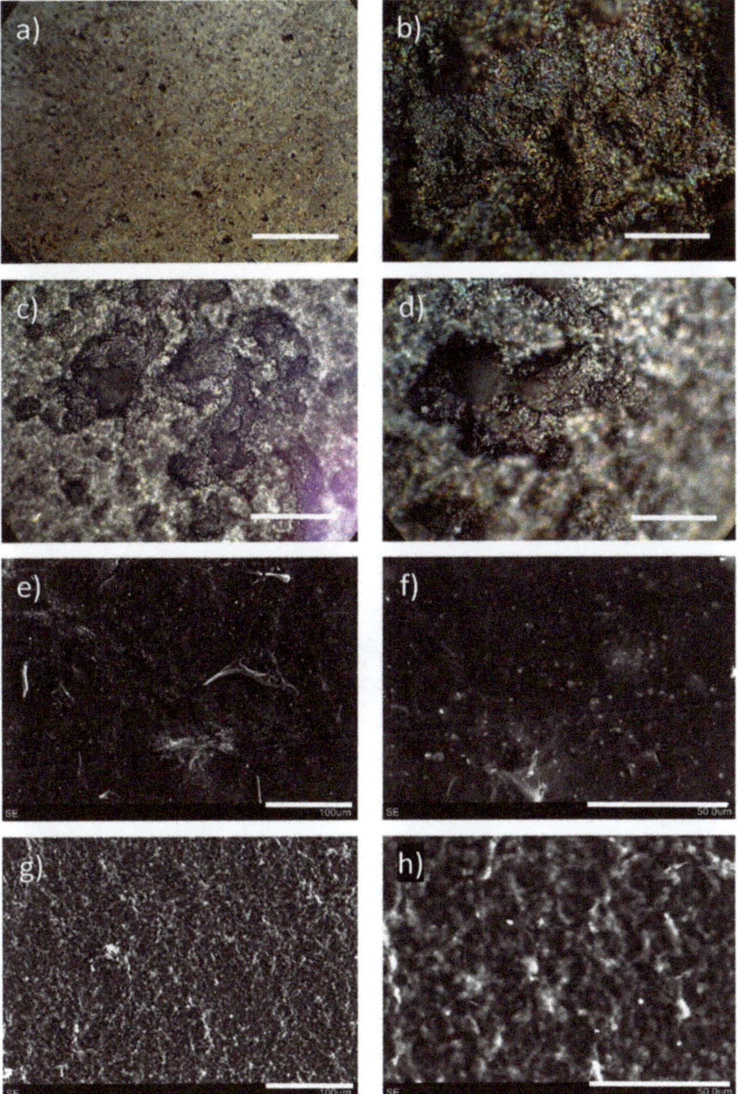

Figure 14. Optical photographs (**a–d**) and SEM images (**e–h**) of films obtained with low-viscosity inks (**a**,**b**,**e**,**f**) and high-viscosity pastes (**c**,**d**,**g**,**h**). Scale bars (in white) = 1 mm (**a**,**b**), 400 μm (**c**,**d**), 100 μm (**e**,**g**), and 50 μm (**f**,**h**). Each image (either optical or from SEM) corresponds to a random point of each sample, none is the direct magnification of another. Used with permission from ref. [70].

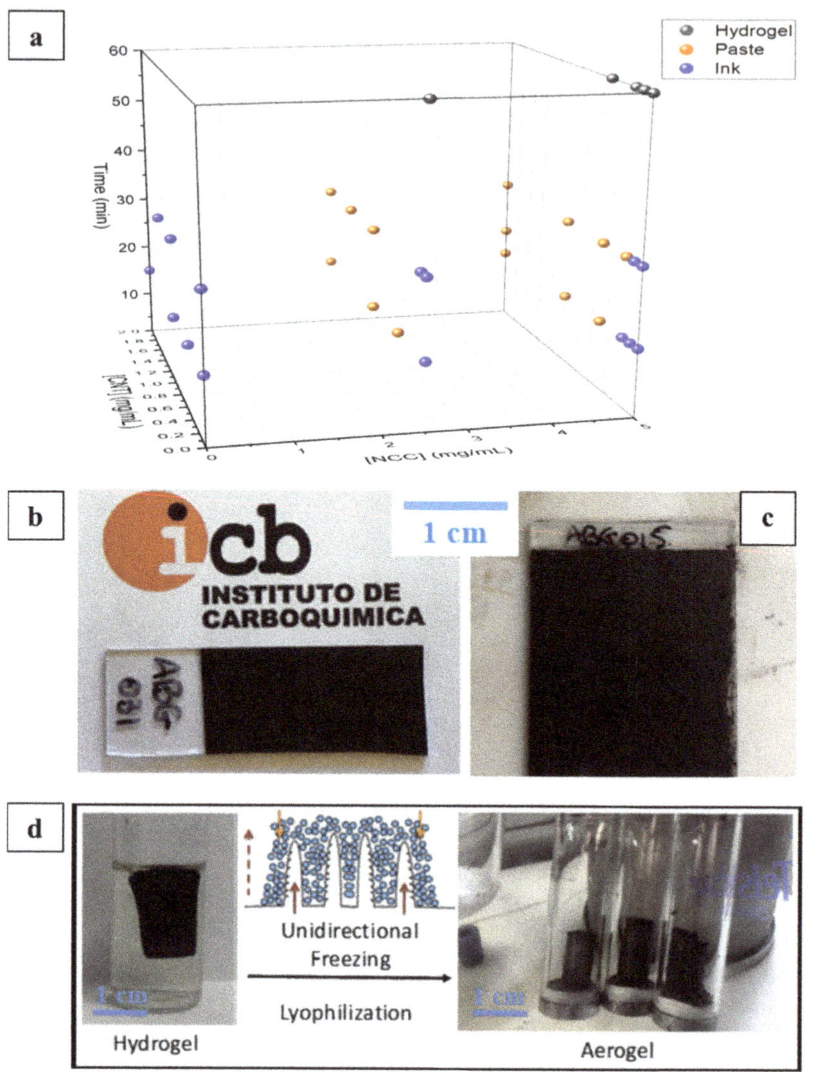

Figure 15. Three-dimensional scatter plot describing the effect of CNTs added in basic pH and conditions to obtain liquid inks (blue), viscous pastes (orange), and self-standing hydrogels (gray) (**a**). All samples had basic pH. Photographs of films made from low-viscosity inks (**b**) and high-viscosity pastes (**c**) on glass substrates. Real images of a hydrogel derived from hydrothermal treatment in water (**left**) and preparation scheme of aerogels by unidirectional freezing followed by lyophilization (**right**) (**d**), used with permission from ref. [70].

Table 1 summarizes examples of 3D printable nanocellulose composites, their sources, formulation, printing method and applications reported in literature.

Table 1. Examples of 3D-printed nanocellulose composites, treatment, and formulation conditions for applications.

Nanocellulose Treatment	Type and Source	Formulation	Printing Method	Area of Application	Reference
Hydrothermal and soda pretreatment, PEG plasticizer, interdigitated electrodes, relative humidity sensing	CNF from bagasse (sugarcane residue)	FS_T3.8_P40; CNF film from bagasse treated with NaOH and 3.8 molar, 40% PEG, IDE screen printing	Screen printing via flatbed screen printer (Everbright S-200HF)	Environmental (relative humidity sensing)	[56]
Carboxymethyl nanocellulose, glycerin, acrylamide, chitosan/AgNPs, cushioning, and antibacterial composite	CMC, (commercially sourced)	(CNGA/C–AgNPs); cushioning 3D-printed structure from 75 vol% glycerine, 1.2 g of N-(2-hydroxyethyl) acrylamide, and functionalized with chitosan–silver nanoparticle (Cts/AgNPS)	Coaxial printing technology via Y&D7300N 3D printer for matrix printing while LSP04-1A syringe pump extrudes the core solution (Cts/AgNPs) concurrently	Food packaging with cushioning and antimicrobial property	[50]
Fused filament fabrication (FFF)-3D printing. Sisal nanocellulose, polylatic acid composite	CNF from raw sisal fibers	(3D-PLA/1CNF); polylactic acid with varying CNF concentrations (1/3/5%)	Compression molding and 3D printing using FFF Desktop 3D printer (Fracktal Works Julia V2)	Diverse	[59]
Hydrothermal treatment, hydrogel–aerogel unidirectional freezing conversion	CNC from cotton-linters-derived microcrystalline cellulose (MCC) powder.	NH_4OH for basic medium, carbon nanotube, and CNC for reinforcement and higher viscosity, reduced Graphene oxide (GO) for conductivity	Spray and rod coating	Metal-free electrodes (electrochemical)	[70]
Plant pulp, nutritional pastes, drying condition	CNF from birch kraft pulp	10% cold swelling starch + 15% SMP, 60% SSMP, 30% rye bran, 35% OPC or 45% FBPC	Extrusion-based printer	Food printing	[65]

4. Conclusions

Three-dimensional printing of nanocellulose-based materials has shown tremendous potential in recent years as evidenced by the increased scientific interest and research work resulting in diverse applications. Recent progress to incorporate 3D printing into industrial applications and the use of nanocellulose as a filler for various composites provides exciting new opportunities for applications due to the excellent characteristics that this material provides, such as enhanced rheological properties, sustainability, biocompatibility, mechanical strength, biodegradability, non-toxicity, and environmental friendliness. This paper discussed the different strategies to produce 3D-printable composites and structures made of nanocellulose with excellent printability and fidelity.

Although a great deal of research has demonstrated the benefits of nanocellulose, studies that explore the printing of these materials in manufacturable products are only in their infancy. Before their potential is realized, more work needs to be done to improve printability and applications. Several future directions could be envisaged. First, there is a need to improve characteristics of nanocellulose-based inks and identify complementary materials and formulations for obtaining suitable printable compositions that result in strong and mechanically stable constructs for applications. A library of tabulated viscosity and rheological properties in relation to the materials' physicochemical properties, the type, content, and stability in the composite mixture, along with the curing conditions, would be beneficial for future developments. Second, more attention needs to be paid to the structural characteristics and understanding of the fundamental changes in the structure and properties of nanocellulose in the composite form upon printing. So far, most optimization on the printability of these composites has been done on a relatively macroscopic scale, and therefore the need for fundamental understanding of these materials that could further accelerate the development of 3D-printable materials is critical for translating this research into new products and applications. Third, the potential of these materials for applications is only beginning to be explored. Further innovations in nanocellulose modification chemistries are needed to add functionality at the surface, which could result in increased mechanical strength, barrier and sorption properties, as well as selectivity for applications such as environmental remediation and sensing. Fourth, there is a need to evaluate the environmental fate and lifecycle of these materials to fully understand their transformation and biodegradability in the environment and/or confirm their biocompatibility upon processing.

In summary, nanocellulose provides innovative solutions to social problems concerning sustainability and moves away from the reliance on petroleum-based products. The recent developments discussed in this review summarize the research status and needs necessary to advance the manufacturability and applications of these promising materials. Future efforts should concentrate on improving printability and understanding the fundamental properties of nanocellulose in printed constructs to enable advanced applications.

Author Contributions: A.S.F., O.P. and S.A. conceived the study; A.S.F. and O.P. performed literature review and prepared the manuscript; S.A. reviewed and completed manuscript. All authors have read and agreed to the published version of the manuscript.

Funding: This research received no external funding.

Conflicts of Interest: The authors declare no conflict of interest.

References

1. Moon, R.J.; Martini, A.; Nairn, J.; Simonsen, J.; Youngblood, J. Cellulose nanomaterials review: Structure, properties and nanocomposites. *Chem. Soc. Rev.* **2011**, *40*, 3941–3994. [CrossRef]
2. Thomas, B.; Raj, M.C.; Athira, K.B.; Rubiyah, M.H.; Joy, J.; Moores, A.; Drisko, G.L.; Sanchez, C. Nanocellulose, a Versatile Green Platform: From Biosources to Materials and Their Applications. *Chem. Rev.* **2018**, *118*, 11575–11625. [CrossRef]
3. Kirk, K.A.; Othman, A.; Andreescu, S. Nanomaterial-functionalized Cellulose: Design, Characterization and Analytical Applications. *Anal. Sci.* **2018**, *34*, 19–31. [CrossRef]
4. Aitomaki, Y.; Oksman, K. Reinforcing efficiency of nanocellulose in polymers. *React. Funct. Polym.* **2014**, *85*, 151–156. [CrossRef]

5. Oksman, K.; Aitomaki, Y.; Mathew, A.P.; Siqueira, G.; Zhou, Q.; Butylina, S.; Tanpichai, S.; Zhou, X.J.; Hooshmand, S. Review of the recent developments in cellulose nanocomposite processing. *Compos. Part A Appl. Sci.* **2016**, *83*, 2–18. [CrossRef]
6. Thunberg, J.; Kalogeropoulos, T.; Kuzmenko, V.; Hagg, D.; Johannesson, S.; Westman, G.; Gatenholm, P. In situ synthesis of conductive polypyrrole on electrospun cellulose nanofibers: Scaffold for neural tissue engineering. *Cellulose* **2015**, *22*, 1459–1467. [CrossRef]
7. Goffin, A.L.; Raquez, J.M.; Duquesne, E.; Siqueira, G.; Habibi, Y.; Dufresne, A.; Dubois, P. Poly(epsilon-caprolactone) based nanocomposites reinforced by surface-grafted cellulose nanowhiskers via extrusion processing: Morphology, rheology, and thermo-mechanical properties. *Polymer* **2011**, *52*, 1532–1538. [CrossRef]
8. Gebhardt, A. *Understanding Additive Manufacturing*; Carl Hanser Verlag GmbH & Co. KG: München, Germany, 2011.
9. Finny, A.S.; Jiang, C.; Andreescu, S. 3D Printed Hydrogel-Based Sensors for Quantifying UV Exposure. *ACS Appl. Mater. Interfaces* **2020**, *12*, 43911–43920. [CrossRef]
10. Othman, A.; Norton, L.; Finny, A.S.; Andreescu, S. Easy-to-use and inexpensive sensors for assessing the quality and traceability of cosmetic antioxidants. *Talanta* **2020**, *208*, 120473. [CrossRef]
11. Abitbol, T.; Rivkin, A.; Cao, Y.F.; Nevo, Y.; Abraham, E.; Ben-Shalom, T.; Lapidot, S.; Shoseyov, O. Nanocellulose, a tiny fiber with huge applications. *Curr. Opin. Biotechnol.* **2016**, *39*, 76–88. [CrossRef] [PubMed]
12. Klemm, D.; Cranston, E.D.; Fischer, D.; Gama, M.; Kedzior, S.A.; Kralisch, D.; Kramer, F.; Kondo, T.; Lindstrom, T.; Nietzsche, S.; et al. Nanocellulose as a natural source for groundbreaking applications in materials science: Today's state. *Mater. Today* **2018**, *21*, 720–748. [CrossRef]
13. Chaichi, M.; Hashemi, M.; Badii, F.; Mohammadi, A. Preparation and characterization of a novel bionanocomposite edible film based on pectin and crystalline nanocellulose. *Carbohydr. Polym.* **2017**, *157*, 167–175. [CrossRef] [PubMed]
14. De France, K.J.; Hoare, T.; Cranston, E.D. Review of Hydrogels and Aerogels Containing Nanocellulose. *Chem. Mater.* **2017**, *29*, 4609–4631. [CrossRef]
15. Norrrahim, M.N.F.; Kasim, N.A.M.; Knight, V.F.; Ujang, F.A.; Janudin, N.; Razak, M.A.I.A.; Shah, N.A.A.; Noor, S.A.M.; Jamal, S.H.; Ong, K.K.; et al. Nanocellulose: The next super versatile material for the military. *Mater. Adv.* **2021**, *2*, 1485–1506. [CrossRef]
16. Chen, W.S.; Yu, H.P.; Lee, S.Y.; Wei, T.; Li, J.; Fan, Z.J. Nanocellulose: A promising nanomaterial for advanced electrochemical energy storage. *Chem. Soc. Rev.* **2018**, *47*, 2837–2872. [CrossRef] [PubMed]
17. Omran, A.A.B.; Mohammed, A.A.B.A.; Sapuan, S.M.; Ilyas, R.A.; Asyraf, M.R.M.; Koloor, S.S.R.; Petru, M. Micro- and Nanocellulose in Polymer Composite Materials: A Review. *Polymers* **2021**, *13*, 231. [CrossRef]
18. Wei, H.; Rodriguez, K.; Renneckar, S.; Vikesland, P.J. Environmental science and engineering applications of nanocellulose-based nanocomposites. *Environ. Sci.-Nano* **2014**, *1*, 302–316. [CrossRef]
19. Azeredo, H.M.C.; Rosa, M.F.; Mattoso, L.H.C. Nanocellulose in bio-based food packaging applications. *Ind. Crop. Prod.* **2017**, *97*, 664–671. [CrossRef]
20. Wang, Y.K.; Wang, L.; Jin, L.; Wang, J.P.; Jin, D.C. Evolutionary Analysis of Cellulose Gene Family in Grasses. *Commun. Comput. Inf. Sci.* **2011**, *244*, 178.
21. Ilyas, R.A.; Sapuan, S.M.; Atiqah, A.; Ibrahim, R.; Abral, H.; Ishak, M.R.; Zainudin, E.S.; Nurazzi, N.M.; Atikah, M.S.N.; Ansari, M.N.M.; et al. Sugar palm (Arenga pinnata [Wurmb.] Merr) starch films containing sugar palm nanofibrillated cellulose as reinforcement: Water barrier properties. *Polym. Compos.* **2020**, *41*, 459–467. [CrossRef]
22. Asyraf, M.R.M.; Ishak, M.R.; Sapuan, S.M.; Yidris, N.; Ilyas, R.A. Woods and composites cantilever beam: A comprehensive review of experimental and numerical creep methodologies. *J. Mater. Res. Technol.* **2020**, *9*, 6759–6776. [CrossRef]
23. Iguchi, M.; Yamanaka, S.; Budhiono, A. Bacterial cellulose—A masterpiece of nature's arts. *J. Mater. Sci.* **2000**, *35*, 261–270. [CrossRef]
24. Dima, S.O.; Panaitescu, D.M.; Orban, C.; Ghiurea, M.; Doncea, S.M.; Fierascu, R.C.; Nistor, C.L.; Alexandrescu, E.; Nicolae, C.A.; Trica, B.; et al. Bacterial Nanocellulose from Side-Streams of Kombucha Beverages Production: Preparation and Physical-Chemical Properties. *Polymers* **2017**, *9*, 374. [CrossRef]
25. Ee, L.Y.; Li, S.F.Y. Recent advances in 3D printing of nanocellulose: Structure, preparation, and application prospects. *Nanoscale Adv.* **2021**, *3*, 1167–1208. [CrossRef]
26. Amin, F.R.; Khalid, H.; Zhang, H.; Rahman, S.U.; Zhang, R.; Liu, G.; Chen, C. Pretreatment methods of lignocellulosic biomass for anaerobic digestion. *AMB Express* **2017**, *7*, 72. [CrossRef] [PubMed]
27. Kumar, A.K.; Sharma, S. Recent updates on different methods of pretreatment of lignocellulosic feedstocks: A review. *Bioresour. Bioprocess.* **2017**, *4*, 7. [CrossRef]
28. Khan, M.N.; Rehman, N.; Sharif, A.; Ahmed, E.; Farooqi, Z.H.; Din, M.I. Environmentally benign extraction of cellulose from dunchi fiber for nanocellulose fabrication. *Int. J. Biol. Macromol.* **2020**, *153*, 72–78. [CrossRef]
29. Zhi, S.L.; Liu, Y.L.; Yu, X.Y.; Wang, X.Y.; Lu, X.B. Enzymatic Hydrolysis of Cellulose after Pretreated by Ionic Liquids: Focus on One-pot Process. *Energy Procedia* **2012**, *14*, 1741–1747. [CrossRef]
30. Smith, E.L.; Abbott, A.P.; Ryder, K.S. Deep Eutectic Solvents (DESs) and Their Applications. *Chem. Rev.* **2014**, *114*, 11060–11082. [CrossRef] [PubMed]
31. Chen, Y.L.; Zhang, X.; You, T.T.; Xu, F. Deep eutectic solvents (DESs) for cellulose dissolution: A mini-review. *Cellulose* **2019**, *26*, 205–213. [CrossRef]

32. Sultan, S.; Abdelhamid, H.N.; Zou, X.D.; Mathew, A.P. CelloMOF: Nanocellulose Enabled 3D Printing of Metal-Organic Frameworks. *Adv. Funct. Mater.* **2019**, *29*, 1805372. [CrossRef]
33. Fortunati, E.; Luzi, F.; Jimenez, A.; Gopakumar, D.A.; Puglia, D.; Thomas, S.; Kenny, J.M.; Chiralt, A.; Torre, L. Revalorization of sunflower stalks as novel sources of cellulose nanofibrils and nanocrystals and their effect on wheat gluten bionanocomposite properties. *Carbohydr. Polym.* **2016**, *149*, 357–368. [CrossRef]
34. Pohanka, M. Photography by Cameras Integrated in Smartphones as a Tool for Analytical Chemistry Represented by an Butyrylcholinesterase Activity Assay. *Sensors* **2015**, *15*, 13752–13762. [CrossRef] [PubMed]
35. Zhang, W.Y.; Yuan, J.Y. Poly(1-Vinyl-1,2,4-triazolium) Poly(Ionic Liquid)s: Synthesis and the Unique Behavior in Loading Metal Ions. *Macromol. Rapid Commun.* **2016**, *37*, 1124–1129. [CrossRef] [PubMed]
36. Zhang, W.; Zhao, Q.; Yuan, J. Porous polyelectrolytes: Charge pores for more functionalities. *Angew. Chem.* **2017**. [CrossRef]
37. Sicard, C.; Glen, C.; Aubie, B.; Wallace, D.; Jahanshahi-Anbuhi, S.; Pennings, K.; Daigger, G.T.; Pelton, R.; Brennan, J.D.; Filipe, C.D.M. Tools for water quality monitoring and mapping using paper-based sensors and cell phones. *Water Res.* **2015**, *70*, 360–369. [CrossRef]
38. Delaney, J.L.; Hogan, C.F.; Tian, J.; Shen, W. Electrogenerated Chemiluminescence Detection in Paper-Based Microfluidic Sensors. *Anal. Chem.* **2011**, *83*, 1300–1306. [CrossRef]
39. Gong, J.; Lin, H.J.; Dunlop, J.W.C.; Yuan, J.Y. Hierarchically Arranged Helical Fiber Actuators Derived from Commercial Cloth. *Adv. Mater.* **2017**, *29*, 1605103. [CrossRef]
40. Lin, H.J.; Gong, J.; Eder, M.; Schuetz, R.; Peng, H.S.; Dunlop, J.W.C.; Yuan, J.Y. Programmable Actuation of Porous Poly(Ionic Liquid) Membranes by Aligned Carbon Nanotubes. *Adv. Mater. Interfaces* **2017**, *4*, 1600768. [CrossRef]
41. Fujisawa, S. Material design of nanocellulose/polymer composites via Pickering emulsion templating. *Polym. J.* **2021**, *53*, 103–109. [CrossRef]
42. Mingwei, T.; Lijun, Q.; Xiansheng, Z.; Kun, Z.; Shifeng, Z.; Xiaoqing, G.; Guangting, H.; Xiaoning, T.; Yaning, S. Enhanced mechanical and thermal properties of regenerated cellulose/graphene composite fibers. *Carbohydr. Polym.* **2014**, *111*, 456–462. [CrossRef]
43. Lasrado, D.; Ahankari, S.; Kar, K. Nanocellulose-based polymer composites for energy applications-A review. *J. Appl. Polym. Sci.* **2020**, *137*, 48959. [CrossRef]
44. Lu, P.; Yang, Y.; Liu, R.; Liu, X.; Ma, J.X.; Wu, M.; Wang, S.F. Preparation of sugarcane bagasse nanocellulose hydrogel as a colourimetric freshness indicator for intelligent food packaging. *Carbohydr. Polym.* **2020**, *249*, 116831. [CrossRef]
45. Li, S.; Liu, S.; Huang, F.; Lin, S.; Zhang, H.; Cao, S.; Chen, L.; He, Z.; Lutes, R.; Yang, J.; et al. Preparation and Characterization of Cellulose-Based Nanofiltration Membranes by Interfacial Polymerization with Piperazine and Trimesoyl Chloride. *ACS Sustain. Chem. Eng.* **2018**, *6*, 13168–13176. [CrossRef]
46. Fujisawa, S.; Togawa, E.; Kuroda, K.; Saito, T.; Isogai, A. Fabrication of ultrathin nanocellulose shells on tough microparticles via an emulsion-templated colloidal assembly: Towards versatile carrier materials. *Nanoscale* **2019**, *11*, 15004–15009. [CrossRef]
47. Elder, B.; Neupane, R.; Tokita, E.; Ghosh, U.; Hales, S.; Kong, Y.L. Nanomaterial Patterning in 3D Printing. *Adv. Mater.* **2020**, *32*, 1907142. [CrossRef] [PubMed]
48. Klar, V.; Pere, J.; Turpeinen, T.; Karki, P.; Orelma, H.; Kuosmanen, P. Shape fidelity and structure of 3D printed high consistency nanocellulose. *Sci. Rep.-UK* **2019**, *9*, 3822. [CrossRef] [PubMed]
49. Markstedt, K.; Escalante, A.; Toriz, G.; Gatenholm, P. Biomimetic Inks Based on Cellulose Nanofibrils and Cross-Linkable Xylans for 3D Printing. *ACS Appl. Mater. Interfaces* **2017**, *9*, 40878–40886. [CrossRef] [PubMed]
50. Zhou, W.; Fang, J.W.; Tang, S.W.; Wu, Z.G.; Wang, X.Y. 3D-Printed Nanocellulose-Based Cushioning-Antibacterial Dual-Function Food Packaging Aerogel. *Molecules* **2021**, *26*, 3543. [CrossRef]
51. Sarwar, M.S.; Niazi, M.B.K.; Jahan, Z.; Ahmad, T.; Hussain, A. Preparation and characterization of PVA/nanocellulose/Ag nanocomposite films for antimicrobial food packaging. *Carbohydr. Polym.* **2018**, *184*, 453–464. [CrossRef]
52. Hartemann, P.; Hoet, P.; Proykova, A.; Fernandes, T.; Baun, A.; De Jong, W.; Filser, J.; Hensten, A.; Kneuer, C.; Maillard, J.Y.; et al. Nanosilver: Safety, health and environmental effects and role in antimicrobial resistance. *Mater. Today* **2015**, *18*, 122–123. [CrossRef]
53. Fourmann, O.; Hausmann, M.K.; Neels, A.; Schubert, M.; Nystrom, G.; Zimmermann, T.; Siqueira, G. 3D printing of shape-morphing and antibacterial anisotropic nanocellulose hydrogels. *Carbohydr. Polym.* **2021**, *259*. [CrossRef]
54. Giubilini, A.; Siqueira, G.; Clemens, F.J.; Sciancalepore, C.; Messori, M.; Nystrom, G.; Bondioli, F. 3D-Printing Nanocellulose-Poly(3-hydroxybutyrate-co-3-hydroxyhexanoate) Biodegradable Composites by Fused Deposition Modeling. *ACS Sustain. Chem. Eng.* **2020**, *8*, 10292–10302. [CrossRef]
55. Salimi, S.; Sotudeh-Gharebagh, R.; Zarghami, R.; Chan, S.Y.; Yuen, K.H. Production of Nanocellulose and Its Applications in Drug Delivery: A Critical Review. *ACS Sustain. Chem. Eng.* **2019**, *7*, 15800–15827. [CrossRef]
56. Syrovy, T.; Maronova, S.; Kubersky, P.; Ehman, N.V.; Vallejos, M.E.; Pretl, S.; Felissia, F.E.; Area, M.C.; Chinga-Carrasco, G. Wide range humidity sensors printed on biocomposite films of cellulose nanofibril and poly(ethylene glycol). *J. Appl. Polym. Sci.* **2019**, *136*, 47920. [CrossRef]
57. Tehrani, Z.; Nordli, H.R.; Pukstad, B.; Gethin, D.T.; Chinga-Carrasco, G. Translucent and ductile nanocellulose-PEG bionanocomposites-A novel substrate with potential to be functionalized by printing for wound dressing applications. *Ind. Crop. Prod.* **2016**, *93*, 193–202. [CrossRef]

58. Sun, F.Z.; Nordli, H.R.; Pukstad, B.; Gamstedt, E.K.; Chinga-Carrasco, G. Mechanical characteristics of nanocellulose-PEG bionanocomposite wound dressings in wet conditions. *J. Mech. Behav. Biomed.* **2017**, *69*, 377–384. [CrossRef]
59. Ambone, T.; Torris, A.; Shanmuganathan, K. Enhancing the mechanical properties of 3D printed polylactic acid using nanocellulose. *Polym. Eng. Sci.* **2020**, *60*, 1842–1855. [CrossRef]
60. Norrrahim, M.N.F.; Kasim, N.A.M.; Knight, V.F.; Misenan, M.S.M.; Janudin, N.; Shah, N.A.A.; Kasim, N.; Yusoff, W.Y.W.; Noor, S.A.M.; Jamal, S.H.; et al. Nanocellulose: A bioadsorbent for chemical contaminant remediation. *RSC Adv.* **2021**, *11*, 7347–7368. [CrossRef]
61. Daochalermwong, A.; Chanka, N.; Songsrirote, K.; Dittanet, P.; Niamnuy, C.; Seubsai, A. Removal of Heavy Metal Ions Using Modified Celluloses Prepared from Pineapple Leaf Fiber. *ACS Omega* **2020**, *5*, 5285–5296. [CrossRef]
62. Mustafa, F.; Andreescu, S. Nanotechnology-based approaches for food sensing and packaging applications. *RSC Adv.* **2020**, *10*, 19309–19336. [CrossRef]
63. Gao, Q.; Lei, M.; Zhou, K.M.; Liu, X.L.; Wang, S.F. Nanocellulose in Food Packaging-A Short Review. *J. Biobased Mater. Bioenergy* **2020**, *14*, 431–443. [CrossRef]
64. Ahankari, S.S.; Subhedar, A.R.; Bhadauria, S.S.; Dufresne, A. Nanocellulose in food packaging: A review. *Carbohydr. Polym.* **2021**, *255*, 117479. [CrossRef] [PubMed]
65. Lille, M.; Nurmela, A.; Nordlund, E.; Metsa-Kortelainen, S.; Sozer, N. Applicability of protein and fiber-rich food materials in extrusion-based 3D printing. *J. Food Eng.* **2018**, *220*, 20–27. [CrossRef]
66. Zhou, R.; Su, S.Q.; Li, Y.F. Effects of cushioning materials on the firmness of Huanghua pears (Pyrus pyrifolia Nakai cv. Huanghua) during distribution and storage. *Packag. Technol. Sci.* **2008**, *21*, 1–11. [CrossRef]
67. Jarimopas, B.; Singh, S.P.; Sayasoonthorn, S.; Singh, J. Comparison of package cushioning materials to protect post-harvest impact damage to apples. *Packag. Technol. Sci.* **2007**, *20*, 315–324. [CrossRef]
68. Lin, M.H.; Chen, J.H.; Chen, F.; Zhu, C.Q.; Wu, D.; Wang, J.; Chen, K.S. Effects of cushioning materials and temperature on quality damage of ripe peaches according to the vibration test. *Food Packag. Shelf* **2020**, *25*. [CrossRef]
69. Karagiannidis, P.G.; Hodge, S.A.; Lombardi, L.; Tomarchio, F.; Decorde, N.; Milana, S.; Goykhman, I.; Su, Y.; Mesite, S.V.; Johnstone, D.N.; et al. Microfluidization of Graphite and Formulation of Graphene-Based Conductive Inks. *ACS Nano* **2017**, *11*, 2742–2755. [CrossRef]
70. Gonzalez-Dominguez, J.M.; Baigorri, A.; Alvarez-Sanchez, M.A.; Colom, E.; Villacampa, B.; Anson-Casaos, A.; Garcia-Bordeje, E.; Benito, A.M.; Maser, W.K. Waterborne Graphene- and Nanocellulose-Based Inks for Functional Conductive Films and 3D Structures. *Nanomaterials* **2021**, *11*, 1435. [CrossRef]
71. Gonzalez-Dominguez, J.M.; Anson-Casaos, A.; Grasa, L.; Abenia, L.; Salvador, A.; Colom, E.; Mesonero, J.E.; Garcia-Bordeje, J.E.; Benito, A.M.; Maser, W.K. Unique Properties and Behavior of Nonmercerized Type-II Cellulose Nanocrystals as Carbon Nanotube Biocompatible Dispersants. *Biomacromolecules* **2019**, *20*, 3147–3160. [CrossRef]

 nanomaterials

MDPI

Article

Date-Palm-Derived Cellulose Nanocrystals as Reinforcing Agents for Poly(vinyl alcohol)/Guar-Gum-Based Phase-Separated Composite Films

Hamid M. Shaikh [1,*], Arfat Anis [1], Anesh Manjaly Poulose [1], Niyaz Ahamad Madhar [2] and Saeed M. Al-Zahrani [1]

1 SABIC Polymer Research Centre, Department of Chemical Engineering, King Saud University, P.O. Box 800, Riyadh 11421, Saudi Arabia; aarfat@ksu.edu.sa (A.A.); apoulose@ksu.edu.sa (A.M.P.); szahrani@ksu.edu.sa (S.M.A.-Z.)
2 Department of Physics and Astronomy, College of Sciences, King Saud University, P.O. Box 2455, Riyadh 11451, Saudi Arabia; nmadhar@ksu.edu.sa
* Correspondence: hamshaikh@ksu.edu.sa

Abstract: The current study delineates the use of date-palm-derived cellulose nanocrystals (dp-CNCs) as reinforcing agents. dp-CNCs were incorporated in varying amounts to poly(vinyl alcohol)/guar-gum-based phase-separated composite films. The films were prepared by using the solution casting method, which employed glutaraldehyde as the crosslinking agent. Subsequently, the films were characterized by bright field and polarizing microscopy, UV-Vis spectroscopy, FTIR spectroscopy, and mechanical study. The microscopic techniques suggested that phase-separated films were formed, whose microstructure could be tailored by incorporating dp-CNCs. At higher levels of dp-CNC content, microcracks could be observed in the films. The transparency of the phase-separated films was not significantly altered when the dp-CNC content was on the lower side. FTIR spectroscopy suggested the presence of hydrogen bonding within the phase-separated films. dp-CNCs showed reinforcing effects at the lowest amount, whereas the mechanical properties of the films were compromised at higher dp-CNC content. Moxifloxacin was included in the films to determine the capability of the films as a drug delivery vehicle. It was found that the release of the drug could be tailored by altering the dp-CNC content within the phase-separated films. In gist, the developed dp-CNC-loaded poly(vinyl alcohol)/guar-gum-based phase-separated composite films could be explored as a drug delivery vehicle.

Keywords: date palm; cellulose nanocrystals; poly(vinyl alcohol); guar gum; phase-separated films; moxifloxacin; drug delivery

Citation: Shaikh, H.M.; Anis, A.; Poulose, A.M.; Madhar, N.A.; Al-Zahrani, S.M. Date-Palm-Derived Cellulose Nanocrystals as Reinforcing Agents for Poly(vinyl alcohol)/Guar-Gum-Based Phase-Separated Composite Films. *Nanomaterials* **2022**, *12*, 1104. https://doi.org/10.3390/nano12071104

Academic Editors: Rajesh Sunasee, Karina Ckless and Linda J. Johnston

Received: 21 February 2022
Accepted: 25 March 2022
Published: 27 March 2022

1. Introduction

Recently, PVA- and polysaccharide-based phase-separated films have been proposed by several authors. Phase-separated systems are a special type of composite system wherein the different phases are formed due to the thermodynamic incompatibility between the two polymeric phases. Yadav et al. (2017) have synthesized PVA and carboxymethyl tamarind gum-based composite films [1]. Herein, the dispersed phase was carboxymethyl tamarind gum, while the PVA formed the continuum matrices. The films were reported to form phase-separated structures. The authors noted that the carboxymethyl tamarind gum-containing composite films supported better human keratinocyte proliferation. The composite films were also found to be capable of antimicrobial drug delivery applications. In another study, the same group further reported that the properties of the composite films were significantly changed when the internal carboxymethyl tamarind gum phase was reinforced with graphene oxide nanosheets [2]. The blank films showed good antimicrobial properties, which was accounted for the antimicrobial activity of graphene oxide nanosheets. The films were also explored for antimicrobial drug delivery applications. It was found that the films were biocompatible towards human keratinocytes.

Polyvinyl alcohol (PVA) is a semi-crystalline polymer [3]. The polymer is highly hydrophilic, due to which it can be easily soluble in water and is an excellent film-forming polymer [4]. It has been extensively used to develop polymeric matrices for wide applications, including drug delivery, tissue engineering, regenerative medicine, food packaging, and sensor development [5,6]. The biological applications of PVA are motivated because of its excellent biocompatibility with human tissue [7]. Nevertheless, the polymer matrices of PVA have been reported to exhibit poor mechanical and thermal stability. As mentioned previously, the hydrophilic nature of PVA results in quick water absorption when placed in the biological environment [8]. The absorption of water molecules results in the mechanical instability of the PVA matrices, thereby leading to their disruption. The polymer matrices of PVA have been crosslinked either by physical or chemical methods for improving mechanical stability even when such matrices are placed in an aqueous environment [9]. Further, various authors have reported the inclusion of reinforcing materials (e.g., nanocellulose, carbon nanotubes, and their derivatives, silver nanoparticles, etc.) for enhancing the mechanical stability of PVA [10–12].

Similarly, many researchers have proposed the synthesis of polymeric architectures by blending PVA and other biological polymers [13,14]. The blending of polymers of biological origin with PVA allows the researchers to modulate the functionality of the polymeric architectures. For example, polymeric structures of PVA and gelatin, alginate, chitosan, or pectin have been proposed for drug delivery and tissue engineering applications [15–18]. Recently, PVA and polysaccharide-based phase-separated films have been proposed by several authors. In such systems, the polysaccharides from the dispersed phase. The main advantage of such a system can be related to the reinforcing effect exerted by the polysaccharide phase, which helps to improve the mechanical stability of the films, compared with the pristine PVA film. However, the mechanical stability of such phase-separated films is composition-dependent. It has been found that, at a critical concentration, the mechanical stability of the phase-separated film is the highest. The alteration in mechanical properties can be related to the microstructural arrangement and the crystallinity of the polymeric phases.

Guar gum (GG) is a widely used polysaccharide-based biopolymer that is extracted from the embryos of *Cyamopsis tetragonoloba* [19]. Chemically, GG has a linear backbone of (1–4)-β-D-mannopyranosyl units, which consists of α-D-galactopyranosyl units as pendants. The gum has been proposed as a coating material for site-specific drug delivery [20]. Similarly, cellulose is another one of the most abundant naturally occurring polysaccharides from plant cell walls in which a glucose unit of β-1,4 D-glucopyranosyl (anhydroglucose (AGU)) is joined linearly in the $^4C_1$-chain configuration. The high degree of polymerization, high surface-area-to-volume ratio, and availability of numerous chemical functional groups in the nanocellulose offer a high loading and binding capacity for drug release [21]. These polysaccharides are extensively used as biomaterials due to their biodegradable and biocompatible characteristics.

Nonetheless, cellulose nanocrystals (CNCs) are also derived from agricultural waste materials. Valorization of the CNCs has been reported to develop several polymeric architectures for various applications, including photonic, pharmaceutical, and biomedical applications. The dp-CNCs are being synthesized from the wastes of the date palm industry in our previous research [22]. Moreover, the reinforcing effect of the dp-CNCs on the phase-separated composite films has not been studied yet. Hence, examining the ability of the dp-CNCs in tailoring the properties of phase-separated composite films seems rather justified.

Therefore, in this study, we propose to develop PVA–guar gum (GG) phase-separated film where the GG would form the dispersed phase, and the PVA would form the external phase. The PVA–GG film would be reinforced with date-palm-derived cellulose nanocrystals (dp-CNCs). A thorough literature survey suggests that composite films of such compositions have not been investigated. Accordingly, PVA–GG (PGC) phase-separated films reinforced with dp-CNCs are fabricated using a solvent casting method, and its structure–property relationship for drug delivery analysis is studied in this study.

2. Materials and Methods

2.1. Materials

Polyvinyl alcohol (Mw 89,000–98,000, 99+% hydrolyzed CAS number 9002-89-5), glutaraldehyde solution 25% (*v*/*v*), were purchased from Sigma Aldrich, St. Louis, MO, USA. Isopropanol (99.9%, analytical grade), and hydrochloric acid (37%, ACS reagent) were obtained from Panreac Química, Garraf, Spain. Guar gum was supplied by Scharlab, Barcelona, Spain. dp-CNCs were synthesized in our lab, as per the method mentioned earlier. In brief, fine powder was obtained from palm tree trunk mesh and pretreated with supercritical carbon dioxide ($ScCO_2$) to eliminate the water-soluble extractives. It was then treated with a 20% (*w*/*v*) sodium hydroxide solution at 90 °C for 6 h. The solid fraction was separated and washed until it became alkali-free. Finally, a bleaching reaction was carried out at 70 °C for 4 h, using an acidic solution of sodium chloride (pH 3.7). The pure cellulose was collected by filtration, washed several times until it became neutral, and finally dried to obtain a constant weight. For obtaining nanocellulose from this cellulose, a combination of mechanical disintegration and chemical treatments with sulfuric acid was used. Firstly, homogenous cellulose suspension has a solid content of ~5 wt.%. The suspension was then fed into the barrel of a twin-screw DSM-Xplore micro-compounder (15 cm^3 Xplore®, Sittard, The Netherland) for mechanical defibrillation of cellulose. It is continuously recirculated within a barrel for about 30 min at a constant screw speed of 250 rpm. After this treatment, solids were collected and subjected to sulfuric acid hydrolysis. In short, 10 g treated cellulose samples were hydrolyzed in 100 mL 50 wt.% H_2SO_4 solution. The reaction was performed at 45 °C with continuous stirring for about 60 min. Finally, the hydrolysis reaction was quenched by the addition of a large amount of distilled water. This suspension was centrifuged several times, and the supernatant fluid was discarded till it became neutral. This cloudy suspension was then dialyzed against distilled water until the pH of the suspension reached a constant value. This portion of the nanocellulose suspension was stored in a refrigerator at 4 °C, while the other was freeze-dried and utilized for further use. Furthermore, it has nanoparticles with sizes ranging from 26 nm to 61 nm, a negative zeta potential of −35 mV, and 89% crystallinity [22]. Furthermore, double distilled water was used throughout the study.

2.2. Preparation of Nanocomposite Films

The PVA–GG-based nanocomposite films were prepared similar to the method proposed by Yadav et al. [2]. In short, initially, solutions of 10% (*w*/*w*) PVA and 2% (*w*/*w*) GG in water were prepared. To make a 10% aqueous solution of PVA, 10 g of PVA was gradually dispersed in 90 g of water by stirring. The mixture was then put in a water bath at 90 °C to achieve complete dissolution of PVA. For GG preparation, 2 wt.% GG was prepared at room temperature by dissolving 2 g GC in 98 g of hot water. Then, the solutions of PVA and GG were mixed in the ratio of 18:2 (*w*/*w*) and homogenized at room temperature using an overhead stirrer (200 rpm; 5 min). A volume of 10 mL of water was then added to the mixture and subsequently homogenized. A suspension of dp-CNC in 10 mL water, containing 0 mg, 2.5 mg, 5 mg, 7.5 mg, and 10 mg, was then added to the diluted solution and homogenized further for another 5 min. This was followed by the addition of the 2 mL of glutaraldehyde reagent, which was used as the crosslinking solution. The glutaraldehyde reagent was prepared by mixing glutaraldehyde, isopropanol, and hydrochloric acid in the ratio of 0.5:0.5:0.05 (i.e., 0.5 mL of glutaraldehyde, 0.5 mL of isopropanol, and 0.05 mL of hydrochloric acid). The liquid mixtures (42 mL) were then converted into films of 60 mm diameter by the solution casting method using 20 mL of the mixtures. During the drying stage, the liquid mixtures, which were poured into Petri dishes, were kept in a thermal cabinet (40 °C) for 24 h. The drying process resulted in the formation of films, which were peeled off with the help of forceps. The films were named A0, A1, A2, A3, and A4, respectively. A0 film was a phase-separated film of PVA and GG, devoid of dp-CNC reinforcement. The rest of the films contained 2.5 mg, 5 mg, 7.5 mg, and 10 mg of dp-CNCs, respectively, in the 40 mL of PVA–GG mixture. The drug-loaded films were synthesized by incorporating 400 mg of moxifloxacin in the liquid mixture after the addition of the crosslinking agent. The films thus formed were named A0D, A1D, A2D, A3D, and A4D. Figure 1 illustrates the flowchart of the film formulation.

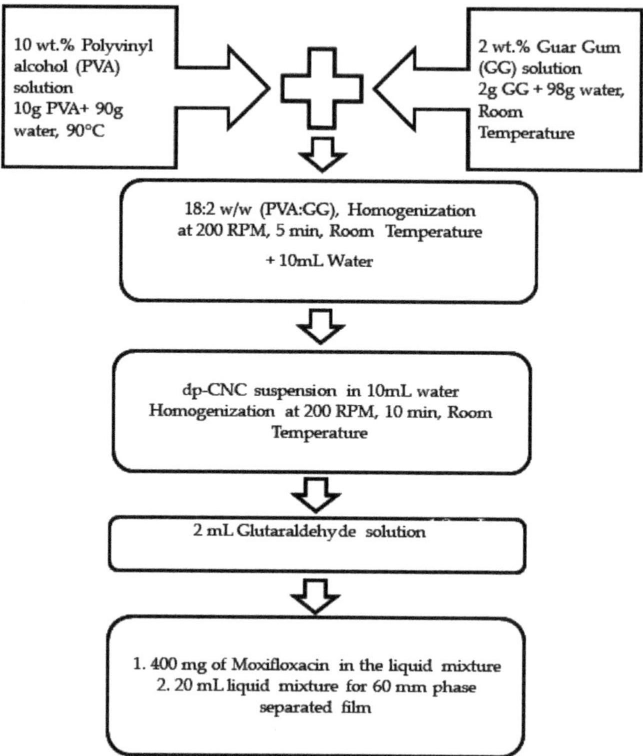

Figure 1. Flowchart diagram of the preparation of the films.

2.3. *Characterization of the Films*

2.3.1. Microstructure Analysis

The microstructures of the films were observed and analyzed under bright-field (Model: DM750, Leica Microsystems, Wetzlar, Germany) and polarizing (Model DM 75, Leica Microsystems, Wetzlar, Germany) microscopes.

2.3.2. Spectroscopic Analyses

The UV-Vis transmission spectrum of the films was evaluated using a UV-Vis spectroscope (Shimadzu 3600 UV-VIS-NIR, Kyoto, Japan). The scanning was carried out in the wavelength region of 280 nm and 800 nm. The functional group analysis and the interactions among the functional groups were analyzed using a Fourier transform infrared (FTIR) spectroscope (Nicolet iN10, Thermo Scientific, Winsford, UK). The measurements were made in the attenuated total reflectance (ATR) mode at room temperature.

2.3.3. X-ray Diffraction Study

The films were subjected to wide-angle X-ray diffraction studies using an X-ray diffractometer (D8 Advance, Bruker, Berlin, Germany). An automated wide-angle goniometer coupled with a sealed tube with Cu-Kα source radiation (λ = 1.54056 Å) was used. In reflection mode, a range of 2θ was scanned from 5° to 50° at 5°/min, and the X-ray tube was operated at 40 kV and 40 mA.

2.3.4. Mechanical Study

The mechanical property of the films was estimated by performing the stress relaxation (SR) study, which can divulge information about the viscoelastic nature of the films. The samples were cut into rectangular pieces (size (L $\times$ B): 60 mm $\times$ 5 mm). Then, the film pieces were attached to the sample holder and placed under the tensile grip. The length of the sample holder window was maintained at 50 mm. The test was carried out by stretching the films by 2 mm and then recording the reduction in the stress values for 60 s. The analysis was carried out in triplicate.

2.3.5. Drug Release Study

The drug release study was conducted in Franz's diffusion cell. The receptor compartment of the diffusion cell was 12.0 mL. The films were cut into circular pieces (1 cm diameter) so that the drug content in the films was 1.57 mg/cm^2. The receptor compartment was filled with phosphate-buffered saline (PBS; 6.8). Then, an activated dialysis tube was placed over the receptor compartment, followed by the placement of the films. Thereafter, the donor compartment was placed and secured, followed by the addition of 1.0 mL of PBS. The PBS in the receptor compartment was sampled (1.00 mL) at regular intervals, which was then replaced with fresh PBS. The sampled PBS was then analyzed in a UV-Vis spectrometer at 391 nm wavelength to determine the drug content. The release study was conducted in triplicate.

2.3.6. Statistical Analysis

The results of mechanical and drug release studies are reported as average $\pm$ standard deviation. The variation in the average values was analyzed by *t*-test.

3. Results and Discussions

3.1. Microstructure Analysis

The addition of dp-CNCs in A1, A2, A3, and A4 was expected to alter the microstructure of the films considerably. The bright-field micrographs (Figure 2) of A1, A2, and A3 showed the presence of the dispersed phases. The agglomeration of the dispersed phase of GG was found to be composition-dependent. Interestingly, it was found that the microarchitecture of A4 was relatively smoother than the other films. Such an observation can only be explained by the ability of dp-CNCs to reduce the interfacial tension among PVA and GG molecules when their concentrations were highest. The bright-field micrographs of A0 film (control), which did not contain dp-CNCs, showed the presence of dispersed phases. This observation was in concurrence with the observations made by Yadav et al. [2], wherein the formation of phase-separated films of PVA was reported.

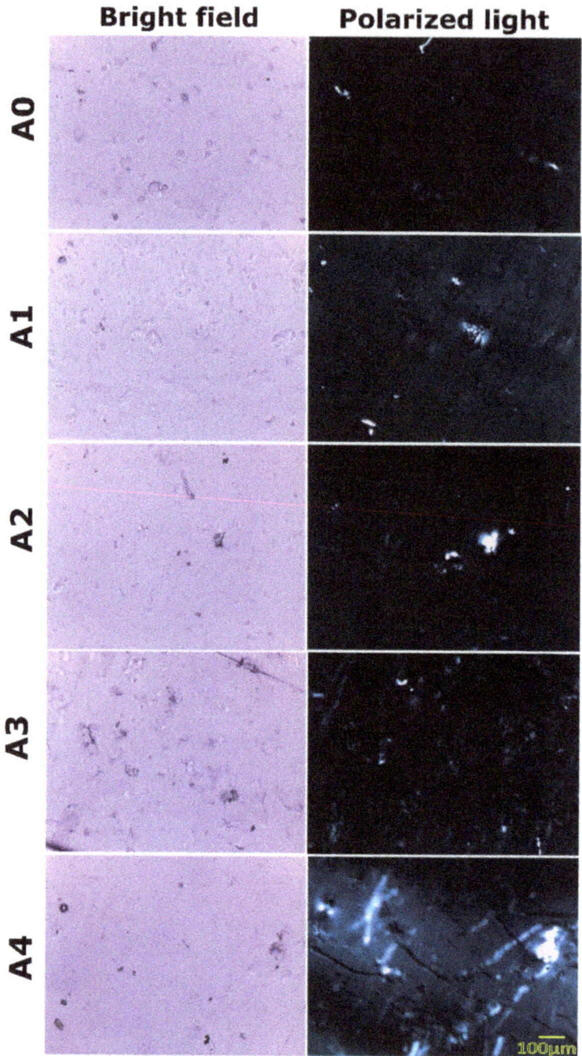

Figure 2. Bright-field and polarizing micrographs of the prepared films.

The polarizing micrographs of A0 also showed the presence of these phase-separated structures. However, the overall brightness of the polarizing micrograph was relatively dark. The dark appearance of the micrographs was due to the positioning of the polarizer and the analyzer in the cross-polarized position. The polarized light micrographs of A1, A2, and A3 confirmed the presence of GG as the dispersed phases. The microarchitecture of A3 showed agglomerated structures of the dispersed phase. The extent of the agglomerated structures was relatively greater in A3 as against A1 and A2, which predominantly showed isolated dispersed phases. Interestingly, the polarized light micrographs of A3 also showed some minor microcracks within the film structure. The polarized light micrographs of A4 showed the extensive presence of microcracks, which were more predominant than the microcracks of A3.

Moreover, the prepared films were colorless and transparent (Figure S1). The formation of colorless and transparent films of PVA and polysaccharides has been reported earlier by several research groups [2]. Such an observation can be related to the presence of PVA, which forms colorless and transparent films, in a higher amount than the polysaccharide. The prepared films were firm to touch and could be appropriately handled without damaging their structure. Apart from this, the foldability of the films was also very good. This can be reasoned to the presence of PVA as the continuum matrix. The films were placed over a scale to judge their apparent transparency (Supplementary Information). It could be observed that the marking of the scale through A0, A1, A2, and A3 could be clearly seen. There was no distortion of the markings on the scale. However, in the case of A4, the markings appeared slightly hazy. This is an indication of reduced transparency when dp-CNC is incorporated within the films in higher quantity.

Nonetheless, researchers are extensively studying the microstructures of polymer matrices. This is because an alteration in the microstructure can alter the properties, including physicochemical and mechanical properties, of the polymer matrices. Even a slight change in the microstructure can greatly alter the properties of the polymer matrices. The polymer scientists have reported adding two or more polymers for tailoring the properties. In this regard, PVA has been blended with various polysaccharides, including GG (pristine or modified), by many researchers [23,24]. In many studies, PVA has been blended together with GG and other polymers (e.g., chitosan, tamarind seed kernel powder, κ-carrageenan, cellulose, etc.) [25–28]. Further, the properties of the blends of PVA and GG have also been tailored using nanoparticles [26]. In all of the papers, it has been reported that the addition of GG, other polymers, and nanoparticles have considerably affected the properties of the PVA films.

3.2. Spectroscopic Analyses

The UV-Vis spectra of the films are shown in Figure 3a. From the spectral profiles, it can be seen that at ~280 nm (start region of UVB radiation), the absorption by all the films was very high, thereby resulting in near-zero transparency. However, by the wavelength of 315 nm (end region of UVB radiation), there was a considerable increase in the transparency values of the films [29]. The A0, A1, A2, and A3 films showed similar transparency values ~73%, but the A4 films showed significantly lower transparency (~62%). In the UVC region, there was an increase in the transparency of the films throughout the radiation wavelengths [30]. The A0, A1, and A2 films showed similar transparency values, higher than A3 and A4 films. The transparency of the A3 film was slightly lower than the transparency of the A0, A1, and A2 films. Nevertheless, the transparency of the A4 films was significantly lower than the rest of the films. Further, in the visible region, the transparency values remained constant. There was not a considerable variation in the transparency values of the films from the transparency values at 400 nm. The spectra in the visible region are concurrent with the results from the apparent transparency. In the apparent transparency test, it was found that there was a considerable distortion in the markings of the scale when overlayed with A4 film.

Figure 3b–f represent the FTIR spectra of the films. The control (A0) film showed the absorption bands at ~2929 cm^{-1}, ~2861 cm^{-1}, ~1649 cm^{-1}, ~1420 cm^{-1}, ~1329 cm^{-1}, ~1086 cm^{-1}, ~1028 cm^{-1}, ~919 cm^{-1}, and ~824 cm^{-1}. The C–H stretching vibrations in the alkyl groups in PVA and guar gum can be attributed to the appearance of the peak at 2929 cm^{-1}. The CH_2 bending vibrations in PVA and guar gum are responsible for the band at 1651 cm^{-1}. The absorption band at 1420.2 cm^{-1}, 1086.2 cm^{-1}, 919.3 cm^{-1}, and 828.6 cm^{-1} were attributed to C–H wagging vibrations in the -CH_2 group, C–O, CH_2 stretching, and C–C stretching vibrations, respectively [31,32]. All the films showed the presence of all the peaks at similar locations. This observation suggested that the interactions in A0 and dp-CNC films were similar. The addition of dp-CNCs in the films did not significantly affect the nature of the interactions.

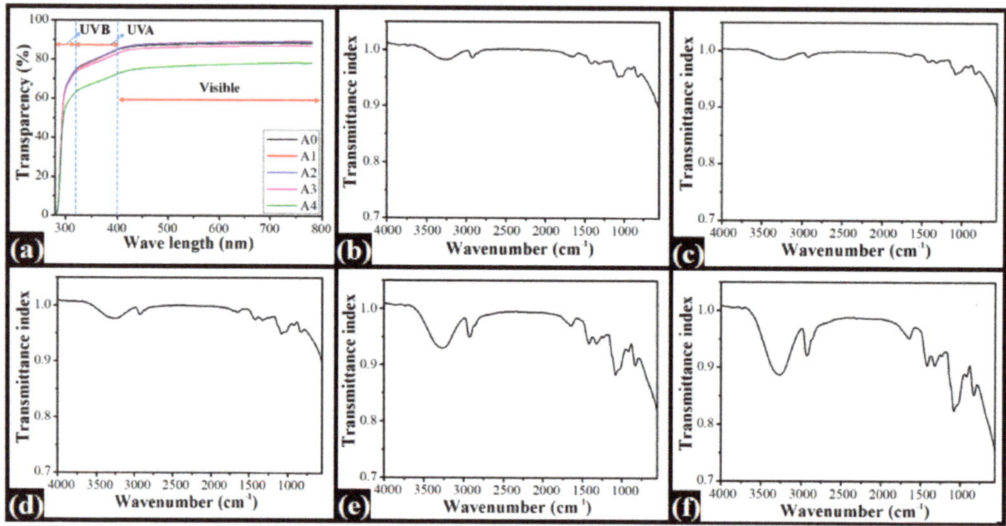

Figure 3. Spectroscopic analyses of the films: (**a**) UV-Vis spectra of the prepared films; (**b–f**) FTIR spectra of the films ((**b**) A0, (**c**) A1, (**d**) A2, (**e**) A3, and (**f**) A4).

Further, there was a broad absorption band in the wavenumber ranges of 3700 cm^{-1} and 2997 cm^{-1} in all of the films (Figure 4). This broad absorption band can be explained by the hydrogen bonding and -OH stretching vibrations within the components of the films [33]. The area under the peak (AUP) of this broad absorption band is a marker of hydrogen bonding. The AUP of the control film (A0) was 7.10. The AUP of A1 film, having the lowest amount of dp-CNCs, decreased to 5.18. However, with a further increase in dp-CNC content within the films A2, A3, and A4, the AUP was increased to 8.61, 21.89, and 33.76, respectively. The increase in AUP could be due to the increase in the intermolecular and intramolecular hydrogen bonds within the film components [34]. A significant increase in the hydrogen bonding may be attributed to the appearance of microcracks in A3 and A4 films.

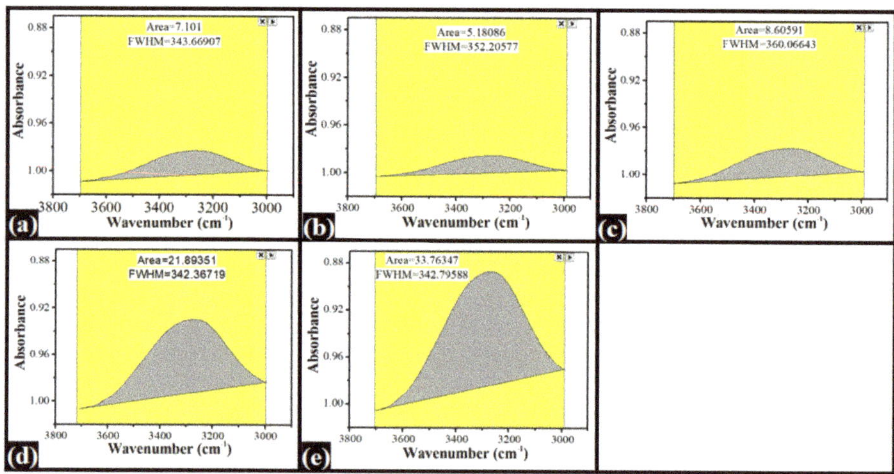

Figure 4. The area under the broad absorption band (3700 cm^{-1} and 2997 cm^{-1}) of the films: (**a**) A0, (**b**) A1, (**c**) A2, (**d**) A3, and (**e**) A4.

3.3. XRD Study

The diffractograms of the films are provided in Figure 5. The control (A0) films showed two major peaks at 12.50° and 20.05° 2θ. The former was a broad peak and had a lower intensity (565.34 arbitrary units (a.u.)). This peak could be associated with the amorphous region within the films. The second peak was a sharp peak, with an intensity of 9271.68 a.u., and could be attributed to the crystalline region. From the intensity values of these peaks, percentage crystallinity (%C) was calculated as per Equation (1) [35]. The %C was found to be 93.90%.

$$\%C = \frac{I_c - I_a}{I_c} \times 100 \quad (1)$$

where %C is the percentage of crystallinity, I_a is the intensity of the amorphous region, and I_c is the intensity of the crystalline region.

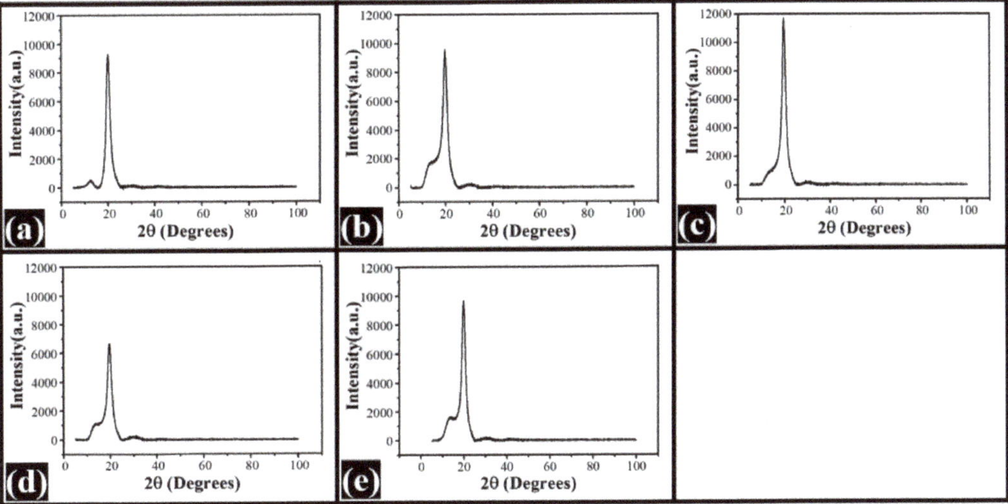

Figure 5. XRD diffractograms of the films: (**a**) A0, (**b**) A1, (**c**) A2, (**d**) A3, and (**e**) A4.

The diffractograms of the dp-CNC-loaded films appeared similar in nature. These films also showed amorphous and crystalline peaks, as seen in A0. However, both of the bands were conjoined, unlike in A0. The amorphous peaks of A1, A2, A3, and A4 were located at 13.64°, 14.7°, 13.85°, and 13.62° 2θ, respectively. It can be observed here that the position of the amorphous peak of A2 was at a higher 2θ value than A1. Thereafter, there was a decrement in the 2θ values in A3 and A4. The corresponding peak intensities were 1787.38, 1026.56, 1133.27, and 1645.09 a.u. As the dp-CNC amount was increased in A2, there was a decrease in the intensity values. In fact, the intensity value of the amorphous peak of A2 was the lowest. After that, the intensity values of the amorphous peaks increased in a concentration-dependent manner. The crystalline peaks in the dp-CNC-loaded films were present at 19.66°, 19.56°, 19.72° and 19.96° 2θ values [35]. The position of the crystalline peak of A2 was at the lowest position. An increment in the dp-CNC content resulted in the increment in the 2θ values of A3 and A4, respectively. The intensity of the crystalline peaks of A1, A2, A3, and A4 were 9575.19, 11,732.9, 6692.58, and 9642.71 a.u., respectively. The results suggested that the highest intensity of the crystalline peak was exhibited by A2. However, no trend regarding the variation in the intensity of the crystalline peaks could be deciphered. %C was calculated from the intensity values of the amorphous and crystalline peaks [35]. The %C of the dp-CNC-loaded films was lower than the control film (93.90%). Among the dp-CNC-loaded films, the %C values of A1, A2,

A3, and A4 were found to be 81.33, 91.25, 83.07, and 82.94, respectively. The %C of A2 was highest, followed by A3, A4, and A1.

3.4. Mechanical Study

The mechanical properties of the films were analyzed by the stress relaxation (SR) study. This method allows the researchers to analyze the changes in the polymer architecture when they are subjected to a constant strain. The SR profiles are shown in Figure 6a. With the increase in the strain, there was an increase in the force values. The force values reached the maximum (F_0), related to firmness, when the films were stretched to the maximum extent (2 mm) (Figure 6b). The F_0 value of A0 was 593.63 ± 43.33 g. The inclusion of dp-CNCs into the films significantly increased the firmness of A1 (F_0 = 908.51 ± 102.26 g), explained by the reinforcing effect of dp-CNC. A further increase in the dp-CNC content correspondingly reduced the F_0 value of the films. The F_0 values of A1 and A2 were significantly different from the control film ($p < 0.05$). The variations in the F_0 values of A3 and A4 were statistically similar to the control film ($p > 0.05$). Among the dp-CNC-loaded films, the F_0 values of A1–A3, A1–A4, and A2–A3 were statistically different ($p < 0.05$).

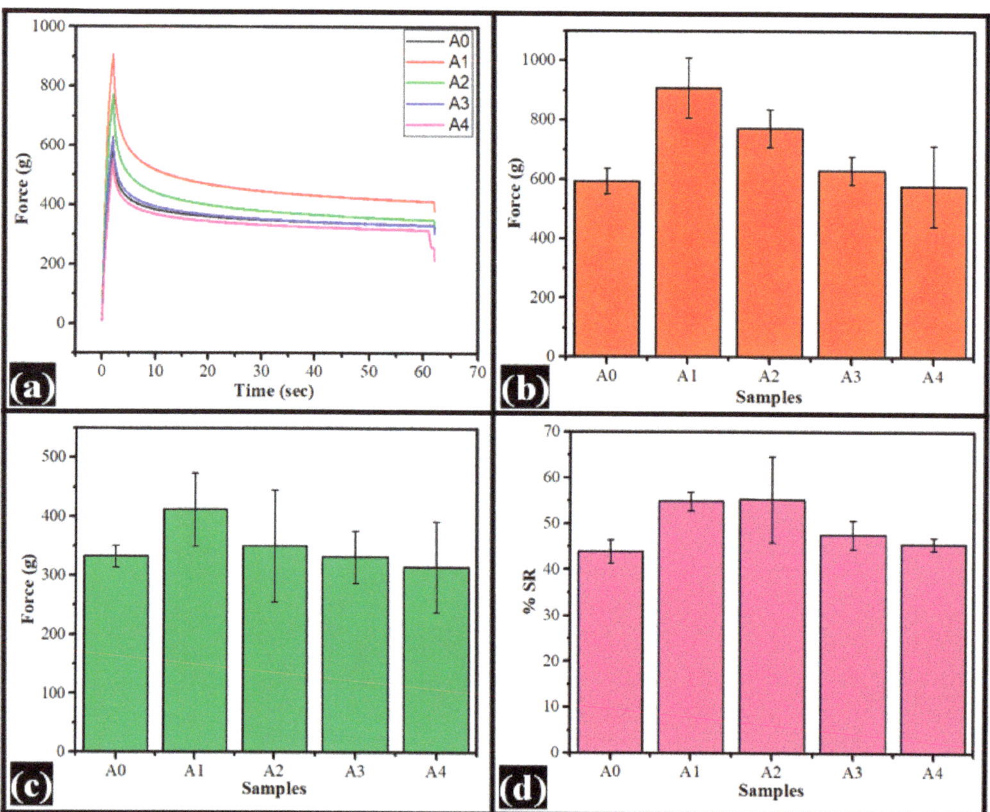

Figure 6. Mechanical properties of the films: (**a**) stress relaxation profiles, (**b**) F_0 values, (**c**) F_{60} values, and (**d**) percent stress relaxation profiles.

The relaxed force values (F_{60}) of the films showed a similar trend as that of F_0 values (Figure 6c). However, the differences in the F_{60} values of the films were statistically insignificant ($p > 0.05$). Subsequently, percentage SR (%SR) values were calculated using the F_0 and F_{60} values (Equation (2)) [36]. The %SR of A_0 was 43.87 ± 2.51 g, which was increased in A1 (54.81 ± 2.09 g) and A2 (55.26 ± 9.42 g) (Figure 6d). The %SR values of A1 and A2 were statistically similar. Even though the %SR values of A0 and A1 were significantly different ($p < 0.05$), the %SR values of A0 and A2 were similar ($p > 0.05$). A consequent increase in the dp-CNCs in the films reduced the %SR values in A3 (47.54 ± 3.08 g) and A4 (45.61 ± 1.40 g), respectively. Among the dp-CNC-containing films, the reduction in the %SR values of A3 and A4 were statistically different ($p < 0.05$). It can be observed that the average %SR values of A1 and A2 were greater than 50%. This is suggestive of a more fluidic component in the films, compared with the others, which are predominantly elastic [37].

$$\%SR = \frac{F_0 - F_{60}}{F_0} \times 100 \tag{2}$$

where %SR is percentage stress relaxation, F_0 is maximum force attained at the maximum strain, and F_{60} is the force at the end of the relaxation period.

In gist, it can be observed that A1 showed higher firmness over the other films due to the reinforcing effect of dp-CNCs. However, it also showed the most increased fluidity, compared with others. Such property can help improve the endurance of the films.

3.5. Drug Release Study

The moxifloxacin drug release profiles from the films are provided in Figure 7. It can be observed that there were corresponding increases in the cumulative percent drug release (CPDR) values as time progressed. However, the CPDR vs. time plot was not a linear plot [29]. An increase in the dp-CNC content in the films correspondingly improved the CPDR values. This may be due to the increase in the hydrophilicity with the increase in the dp-CNC content. The CPDR value of A0D, A1D, A2D, A3D, and A4D at the end of the study was 31.36 ± 2.15%, 33.17 ± 0.35%, 35.56 ± 1.31%, 42.60 ± 1.01%, and 45.45 ± 1.01%, respectively. Even though the average CPDR values of A1D and A2D were higher than the CPDR value of A0D, the values were not statistically significant ($p > 0.05$). The CPDR values of A3D and A4D were significantly higher than A0D ($p < 0.05$). Among the dp-CNC containing films, except the CPDR values of A1D and A2D that were similarly valued ($p > 0.05$), all other CPDR values were significantly different ($p < 0.05$). In gist, the release of the drug from the films was not affected significantly, compared with the control, in which the dp-CNC content was on a lower side. Nevertheless, when the dp-CNC content was on a higher side, the release of the drug was considerably increased.

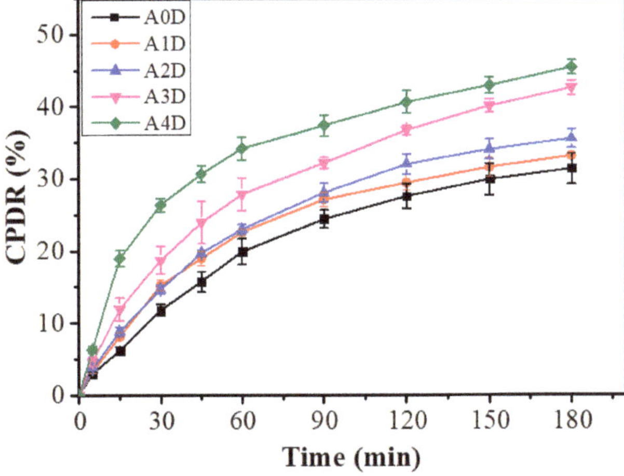

Figure 7. Cumulative percent drug release profiles: A0D, without dp-CNC, with 400 mg of drug; A1D, 2.5 mg dp-CNC, with 400 mg of drug; A2D, 5 mg dp-CNC, with 400 mg of drug; A3D, 7.5 mg dp-CNC, with 400 mg of drug; A4D, 10 mg of dp-CNC, with 400 mg of drug.

4. Conclusions

In the current study, the synthesis of poly(vinyl alcohol)/guar-gum-based phase-separated film was discussed, which was confirmed by the bright-field and polarizing microscopy techniques. The phase-separated film thus formed was incorporated with dp-CNC as the reinforcing agent. The micrographs suggested the presence of two distinct phases within the films. The appearance of such microarchitectures can be attributed to the formation of phase-separated films. It was found that at the lowest dp-CNC content, dp-CNCs acted as reinforcing agents. Interestingly, the hydrogen bonding was lowest in this film. An increase in the dp-CNC content significantly improved the hydrogen bonding. However, the mechanical properties of the films were compromised significantly in A2, A3, and A4, compared with A1. In fact, in A3 and A4, the hydrogen bonds were so strong that they formed microcracks. These microcracks can be explained by the decrease in the mechanical properties of the A3 and A4 films. Lastly, it was found that the developed films were suitable as delivery vehicles for Moxifloxacin. In the future, research with other types of drugs will be performed. Additionally, the biocompatibility of the films will be ascertained using in vitro cell culture studies and in vivo animal studies.

Supplementary Materials: The following supporting information can be downloaded at: https://www.mdpi.com/article/10.3390/nano12071104/s1, Figure S1: The prepared phase-separated films placed over scales to analyze morphology and transparency.

Author Contributions: Conceptualization, H.M.S. and A.A.; methodology, H.M.S. and A.A.; investigation, H.M.S., A.M.P. and N.A.M.; resources, S.M.A.-Z. and H.M.S., writing—original draft preparation, H.M.S.; writing—review and editing, H.M.S. and A.A.; supervision, S.M.A.-Z.; funding acquisition, H.M.S. All authors have read and agreed to the published version of the manuscript.

Funding: This project was funded by the National Plan for Science, Technology, and Innovation (MAARIFAH), King Abdulaziz City for Science and Technology, Kingdom of Saudi Arabia, Award Number (13-NAN1120-02).

Institutional Review Board Statement: Not applicable.

Informed Consent Statement: Not applicable.

Data Availability Statement: Data are contained within the article.

Conflicts of Interest: The authors declare no conflict of interest.

References

1. Yadav, I.; Rathnam, V.S.S.; Yogalakshmi, Y.; Chakraborty, S.; Banerjee, I.; Anis, A.; Pal, K. Synthesis and characterization of polyvinyl alcohol-carboxymethyl tamarind gum based composite films. *Carbohydr. Polym.* **2017**, *165*, 159–168. [CrossRef] [PubMed]
2. Yadav, I.; Nayak, S.K.; Rathnam, V.S.S.; Banerjee, I.; Ray, S.S.; Anis, A.; Pal, K. Reinforcing effect of graphene oxide reinforcement on the properties of poly (vinyl alcohol) and carboxymethyl tamarind gum based phase-separated film. *J. Mech. Behav. Biomed. Mater.* **2018**, *81*, 61–71. [CrossRef] [PubMed]
3. Ali, H.E.; Algarni, H.; Khairy, Y. Influence of cobalt-metal concentration on the microstructure and optical limiting properties of PVA. *Opt. Mater.* **2020**, *108*, 110212. [CrossRef]
4. Arefian, M.; Hojjati, M.; Tajzad, I.; Mokhtarzade, A.; Mazhar, M.; Jamavari, A. A review of Polyvinyl alcohol/Carboxymethyl cellulose (PVA/CMC) composites for various applications. *J. Compos. Compd.* **2020**, *2*, 69–76.
5. Zulkiflee, I.; Fauzi, M.B. Gelatin-Polyvinyl Alcohol Film for Tissue Engineering: A Concise Review. *Biomedicines* **2021**, *9*, 979. [CrossRef]
6. Rezaei, F.S.; Sharifianjazi, F.; Esmaeilkhanian, A.; Salehi, E. Chitosan films and scaffolds for regenerative medicine applications: A review. *Carbohydr. Polym.* **2021**, *273*, 118631. [CrossRef]
7. Bi, S.; Wang, P.; Hu, S.; Li, S.; Pang, J.; Zhou, Z.; Sun, G.; Huang, L.; Cheng, X.; Xing, S. Construction of physical-crosslink chitosan/PVA double-network hydrogel with surface mineralization for bone repair. *Carbohydr. Polym.* **2019**, *224*, 115176. [CrossRef]
8. Feng, R.; Fu, R.; Duan, Z.; Zhu, C.; Ma, X.; Fan, D.; Li, X. Preparation of sponge-like macroporous PVA hydrogels via n-HA enhanced phase separation and their potential as wound dressing. *J. Biomater. Sci. Polym. Ed.* **2018**, *29*, 1463–1481. [CrossRef]
9. Gadhave, R.V.; Mahanwar, P.A.; Gadekar, P.T. Effect of glutaraldehyde on thermal and mechanical properties of starch and polyvinyl alcohol blends. *Des. Monomers Polym.* **2019**, *22*, 164–170. [CrossRef]

10. Sarwar, M.S.; Niazi, M.B.K.; Jahan, Z.; Ahmad, T.; Hussain, A. Preparation and characterization of PVA/nanocellulose/Ag nanocomposite films for antimicrobial food packaging. *Carbohydr. Polym.* **2018**, *184*, 453–464. [CrossRef]
11. Zhang, H.; Zhang, J. The preparation of novel polyvinyl alcohol (PVA)-based nanoparticle/carbon nanotubes (PNP/CNTs) aerogel for solvents adsorption application. *J. Colloid Interface Sci.* **2020**, *569*, 254–266. [CrossRef] [PubMed]
12. Rolim, W.R.; Pieretti, J.C.; Renó, D.B.L.; Lima, B.A.; Nascimento, M.N.H.; Ambrosio, F.N.; Lombello, C.B.; Brocchi, M.; de Souza, A.C.S.; Seabra, A.B. Antimicrobial activity and cytotoxicity to tumor cells of nitric oxide donor and silver nanoparticles containing PVA/PEG films for topical applications. *ACS Appl. Mater. Interfaces* **2019**, *11*, 6589–6604. [CrossRef] [PubMed]
13. Maheshwari, T.; Tamilarasan, K.; Selvasekarapandian, S.; Chitra, R.; Kiruthika, S. Investigation of blend biopolymer electrolytes based on Dextran-PVA with ammonium thiocyanate. *J. Solid State Electrochem.* **2021**, *25*, 755–765. [CrossRef]
14. Samadi, N.; Sabzi, M.; Babaahmadi, M. Self-healing and tough hydrogels with physically cross-linked triple networks based on Agar/PVA/Graphene. *Int. J. Biol. Macromol.* **2018**, *107*, 2291–2297. [CrossRef]
15. Perez-Puyana, V.; Jiménez-Rosado, M.; Romero, A.; Guerrero, A. Development of PVA/gelatin nanofibrous scaffolds for Tissue Engineering via electrospinning. *Mater. Res. Express* **2018**, *5*, 035401. [CrossRef]
16. Jadbabaei, S.; Kolahdoozan, M.; Naeimi, F.; Ebadi-Dehaghani, H. Preparation and characterization of sodium alginate–PVA polymeric scaffolds by electrospinning method for skin tissue engineering applications. *RSC Adv.* **2021**, *11*, 30674–30688. [CrossRef]
17. Mombini, S.; Mohammadnejad, J.; Bakhshandeh, B.; Narmani, A.; Nourmohammadi, J.; Vahdat, S.; Zirak, S. Chitosan-PVA-CNT nanofibers as electrically conductive scaffolds for cardiovascular tissue engineering. *Int. J. Biol. Macromol.* **2019**, *140*, 278–287. [CrossRef]
18. Kraskouski, A.; Hileuskaya, K.; Kulikouskaya, V.; Kabanava, V.; Agabekov, V.; Pinchuk, S.; Vasilevich, I.; Volotovski, I.; Kuznetsova, T.; Lapitskaya, V. Polyvinyl alcohol and pectin blended films: Preparation, characterization, and mesenchymal stem cells attachment. *J. Biomed. Mater. Res. Part A* **2021**, *109*, 1379–1392. [CrossRef]
19. Verma, D.; Sharma, S.K. Recent advances in guar gum based drug delivery systems and their administrative routes. *Int. J. Biol. Macromol.* **2021**, *181*, 653–671. [CrossRef]
20. Kumar, B.; Murali, A.; Bharath, A.B.; Giri, S. Guar gum modified upconversion nanocomposites for colorectal cancer treatment through enzyme-responsive drug release and NIR-triggered photodynamic therapy. *Nanotechnology* **2019**, *30*, 315102. [CrossRef]
21. Das, S.; Ghosh, B.; Sarkar, K. Nanocellulose as sustainable biomaterials for drug delivery. *Sens. Int.* **2022**, *3*, 100135. [CrossRef]
22. Shaikh, H.M.; Anis, A.; Poulose, A.M.; Al-Zahrani, S.M.; Madhar, N.A.; Alhamidi, A.; Alam, M.A. Isolation and Characterization of Alpha and Nanocrystalline Cellulose from Date Palm (Phoenix dactylifera L.) Trunk Mesh. *Polymers* **2021**, *13*, 1893. [CrossRef] [PubMed]
23. Dalei, G.; Das, S.; Das, S.P. Low-pressure nitrogen and ammonia plasma treatment on carboxymethyl guar gum/PVA hydrogels: Impact on drug delivery, biocompatibility and biodegradability. *Int. J. Polym. Mater. Polym. Biomater.* **2021**, *70*, 75–89. [CrossRef]
24. Gupta, A.P.; Arora, G. Preparation and characterization of guar-gum/polyvinylalcohol blend films. *J. Mater. Sci. Eng. B* **2011**, *1*, 28.
25. Iqbal, D.N.; Shafiq, S.; Khan, S.M.; Ibrahim, S.M.; Abubshait, S.A.; Nazir, A.; Abbas, M.; Iqbal, M. Novel chitosan/guar gum/PVA hydrogel: Preparation, characterization and antimicrobial activity evaluation. *Int. J. Biol. Macromol.* **2020**, *164*, 499–509. [CrossRef]
26. Gasti, T.; Hiremani, V.D.; Kesti, S.S.; Vanjeri, V.N.; Goudar, N.; Masti, S.P.; Thimmappa, S.C.; Chougale, R.B. Physicochemical and Antibacterial Evaluation of Poly (Vinyl Alcohol)/Guar Gum/Silver Nanocomposite Films for Food Packaging Applications. *J. Polym. Environ.* **2021**, *29*, 3347–3363. [CrossRef]
27. Iqbal, D.N.; Tariq, M.; Khan, S.M.; Gull, N.; Iqbal, S.S.; Aziz, A.; Nazir, A.; Iqbal, M. Synthesis and characterization of chitosan and guar gum based ternary blends with polyvinyl alcohol. *Int. J. Biol. Macromol.* **2020**, *143*, 546–554. [CrossRef]
28. Patil, C.M.; Borse, A.U.; Meshram, J.S. Ionic liquid: Green solvent for the synthesis of cellulose/guar gum/PVA biocomposite. *Green Mater.* **2018**, *6*, 23–29. [CrossRef]
29. Qureshi, D.; Sahoo, A.; Mohanty, B.; Anis, A.; Kulikouskaya, V.; Hileuskaya, K.; Agabekov, V.; Sarkar, P.; Ray, S.S.; Maji, S.; et al. Fabrication and Characterization of Poly (vinyl alcohol) and Chitosan Oligosaccharide-Based Blend Films. *Gels* **2021**, *7*, 55. [CrossRef]
30. Qureshi, D.; Pattanaik, S.; Mohanty, B.; Anis, A.; Kulikouskaya, V.; Hileuskaya, K.; Agabekov, V.; Sarkar, P.; Maji, S.; Pal, K. Preparation of novel poly(vinyl alcohol)/chitosan lactate-based phase-separated composite films for UV-shielding and drug delivery applications. *Polym. Bull.* **2021**. [CrossRef]
31. Mathew, S.; Jayakumar, A.; Kumar, V.P.; Mathew, J.; Radhakrishnan, E.K. One-step synthesis of eco-friendly boiled rice starch blended polyvinyl alcohol bionanocomposite films decorated with in situ generated silver nanoparticles for food packaging purpose. *Int. J. Biol. Macromol.* **2019**, *139*, 475–485. [CrossRef] [PubMed]
32. Singh, S.; Gaikwad, K.K.; Lee, Y.S. Antimicrobial and antioxidant properties of polyvinyl alcohol bio composite films containing seaweed extracted cellulose nano-crystal and basil leaves extract. *Int. J. Biol. Macromol.* **2018**, *107*, 1879–1887. [CrossRef] [PubMed]
33. Hendrawan, H.; Khoerunnisa, F.; Sonjaya, Y.; Putri, A.D. Poly (vinyl alcohol)/glutaraldehyde/Premna oblongifolia merr extract hydrogel for controlled-release and water absorption application. In Proceedings of the IOP Conference Series: Materials Science and Engineering, Semarang, Indonesia, 7–8 September 2018; p. 012048.
34. Nizan, N.S.N.H.; Zulkifli, F.H. Reinforcement of hydroxyethyl cellulose/poly (vinyl alcohol) with cellulose nanocrystal as a bone tissue engineering scaffold. *J. Polym. Res.* **2020**, *27*, 1–9. [CrossRef]

35. Sagiri, S.S.; Singh, V.K.; Pal, K.; Banerjee, I.; Basak, P. Stearic acid based oleogels: A study on the molecular, thermal and mechanical properties. *Mater. Sci. Eng. C* **2015**, *48*, 688–699. [CrossRef] [PubMed]
36. Das, P.; Qureshi, D.; Paul, S.; Mohanty, B.; Anis, A.; Verma, S.; Wilczyński, S.; Pal, K. Effect of sorbitan monopalmitate on the polymorphic transitions and physicochemical properties of mango butter. *Food Chem.* **2021**, *347*, 128987. [CrossRef] [PubMed]
37. Bharti, D.; Kim, D.; Cerqueira, M.A.; Mohanty, B.; Habibullah, S.; Banerjee, I.; Pal, K. Effect of Biodegradable Hydrophilic and Hydrophobic Emulsifiers on the Oleogels Containing Sunflower Wax and Sunflower Oil. *Gels* **2021**, *7*, 133. [CrossRef] [PubMed]

 *nanomaterials*

Article

Vascular and Blood Compatibility of Engineered Cationic Cellulose Nanocrystals in Cell-Based Assays

Alexandre Bernier, Tanner Tobias, Hoang Nguyen, Shreshth Kumar, Beza Tuga, Yusha Imtiaz, Christopher W. Smith, Rajesh Sunasee * and Karina Ckless *

Department of Chemistry, State University of New York at Plattsburgh, Plattsburgh, NY 12901, USA; abern012@plattsburgh.edu (A.B.); ttobi004@plattsburgh.edu (T.T.); hnguy021@plattsburgh.edu (H.N.); skuma010@plattsburgh.edu (S.K.); tuga0002@umn.edu (B.T.); yusha.imtiaz@emory.edu (Y.I.); C.W.Smith022@gmail.com (C.W.S.)
* Correspondence: rajesh.sunasee@plattsburgh.edu (R.S.); kckle001@plattsburgh.edu (K.C.)

Abstract: An emerging interest regarding nanoparticles (NPs) concerns their potential immunomodulatory and pro-inflammatory activities, as well as their impact in the circulatory system. These biological activities of NPs can be related to the intensity and type of the responses, which can raise concerns about adverse side effects and limit the biomedical applicability of these nanomaterials. Therefore, the purpose of this study was to investigate the impact of a library of cationic cellulose nanocrystals (CNCs) in the human blood and endothelial cells using cell-based assays. First, we evaluated whether the cationic CNCs would cause hemolysis and aggregation or alteration on the morphology of red blood cells (RBC). We observed that although these nanomaterials did not alter RBC morphology or cause aggregation, at 24 h exposure, a mild hemolysis was detected mainly with unmodified CNCs. Then, we analyzed the effect of various concentrations of CNCs on the cell viability of human umbilical vein endothelial cells (HUVECs) in a time-dependent manner. None of the cationic CNCs caused a dose-response decrease in the cell viability of HUVEC at 24 h or 48 h of exposure. The findings of this study, together with the immunomodulatory properties of these cationic CNCs previously published, support the development of engineered cationic CNCs for biomedical applications, in particular as vaccine nanoadjuvants.

Keywords: cellulose nanocrystals; cationic; immunomodulator; hemolysis; cytotoxicity

Citation: Bernier, A.; Tobias, T.; Nguyen, H.; Kumar, S.; Tuga, B.; Imtiaz, Y.; Smith, C.W.; Sunasee, R.; Ckless, K. Vascular and Blood Compatibility of Engineered Cationic Cellulose Nanocrystals in Cell-Based Assays. *Nanomaterials* **2021**, *11*, 2072. https://doi.org/10.3390/nano11082072

Academic Editor: Takuya Kitaoka

Received: 12 July 2021
Accepted: 13 August 2021
Published: 15 August 2021

1. Introduction

The medical nanotechnology field has grown exponentially in the past 10 years and the possibilities of applications of cellulose nanocrystals (CNCs) in this context have been expanded from the initial proposed use as drug delivery platforms [1] to sophisticated bio-imaging [2] and pH sensing [3] systems, among others [4]. The source of CNCs is cellulose. This polymer, made of glucose, plays a crucial role in maintaining the structure of the plant cell wall and it is the most abundant polysaccharide on earth [5–9]. CNCs are unique nanomaterial obtained from the acid hydrolysis of native cellulose, forming rigid "rod-like" crystalline nanocellulose (length typically between 100–200 nm and diameter ~5–10 nm). They exhibit remarkable strength and physicochemical properties including a high aspect ratio, low density, and large specific surface area, as well as have the presence of abundant hydroxyl groups for surface chemical modifications [7,10,11]. In fact, the presence of abundant hydroxyl groups allows these nanomaterials to be engineered by modifying these groups with various functional polymers and pendants with the purpose to be tailored for specific biomedical applications [9], including potential immunomodulators. Polysaccharides, such as 2,3-O-acetylated-1,4-β-D-glucomannan, have been shown to elicit an immune response by stimulating the secretion of cytokines, interleukin 1-beta (IL-1β), and tumor necrosis factor alpha (TNF-α) in human cell lines [12]. The immune response is a physiological event that occurs in many biological systems and it is the foundation of vaccine effectiveness. The proper immune response and lasting immunity towards a pathogen elicited by a vaccine is sometimes achieved only in the presence of adjuvants, which are substances that help to boost this response [13–15]. Adjuvants currently approved as components of vaccines are particulate nanomaterials such as "alum" (aluminum oxohydroxide and aluminum hydroxyphosphate) or oil-in-water emulsions, which act as both vaccine delivery vehicles and immunostimulants [16,17]. The emerging interest in CNCs for biomedical applications as well as their particulate morphology prompted us to develop a series of wood-based CNCs with positive surface charges and potential immunomodulatory activities that hopefully could be further developed in newly engineered vaccine nanoadjuvants. In fact, the immunostimulatory activity of cationic CNCs was first described in our previous work in which we found that they induced the secretion of the inflammatory cytokine, IL-1β, in mouse and human macrophage cells [18,19]. Recently, we successfully expanded the library of cationic CNCs by engineering the surface of CNCs with poly[2-(methacryloyloxy)ethyl]trimethylammonium chloride (METAC) and poly(aminoethyl methacrylate hydrochloride (AEM)) possessing pendant cationic groups (+NMe_3 and +NH_3 respectively)[20]. By changing the proportion of initiators (2-bromoisobutyryl bromide, Brib) and monomers (METAC and AEM) during the polymerization process, we obtained a series of cationic CNCs with different surface charges and hydrodynamic sizes. This new library of cationic CNCs was evaluated using three different cell-based assays and relevant immune cells, including mouse cell lines and human peripheral blood mononuclear cells (PBMCs). Overall, we demonstrated that these cellulose-based nanomaterials present very low cytotoxicity in all our tested experimental conditions [20]. Given that these cationic cellulose-based nanomaterials have the potential to be developed as immunomodulators and therefore vaccine nanoadjuvants, it is paramount that all aspects of their interactions with biological systems must be evaluated. The majority of currently utilized vaccines are administered intramuscularly (i.e., direct injection into the skeletal muscle) [15], implying that these nanoadjuvants have the possibility to be in contact with blood and vascular cells. As part of the biomedical application safety assessment of nanoparticles, the blood compatibility assays comprehend a series of tests to verify the interaction between nanomaterials and blood components alongside its consequences including hemolysis [21] and RBC aggregation and morphology [22]. Thus, in this study, we utilized RBC lysis, aggregation, and morphology, as well as cytotoxicity in endothelial cells in dose-response and time course studies to evaluate the blood and vascular compatibility of engineered cationic CNCs. Overall, we demonstrated that unmodified and modified CNCs are compatible with RBC and endothelial cells, and therefore can be further investigated as potential vaccine nanoadjuvants.

2. Materials and Methods

2.1. Cationic CNCs and Preparation of Their Colloidal Suspensions for Biological Assays

The unmodified CNCs used in this study were spray-dried CNCs obtained via sulfuric acid hydrolysis of hardwood pulp that were kindly supplied by InnoTech Alberta Inc. (Edmonton, AB, Canada). The unmodified CNCs possess a negative surface charge due to the presence of the sulfate half-ester groups. CNCs conjugated with the cationic polymers [2-(methacryloyloxy)ethyl]trimethylammonium chloride (METAC) and 2-aminoethyl methacrylate hydrochloride (AEM) were synthesized via surface-initiated single electron transfer living radical polymerization and characterized using analytical, spectroscopy and microscopy techniques as described in our recent publication [20]. The chemical structures of unmodified CNCs and engineered cationic CNCs (also referred to as modified CNCs) are depicted in Scheme 1 and their respective compositions, apparent particle sizes, and zeta potentials are illustrated in Table S1. The cationic CNCs used in this study are CNC-AEM-1A, CNC-AEM-2A, CNC-METAC-1A, CNC-METAC-2A, and CNC-METAC-2B. While CNC-METAC (1A, 1B, and 2A) have the same chemical structures, they differ in terms of their composition based on the amount of initiator and monomer used during the polymerization. For instance, CNC-METAC-1A and CNC-METAC-1B were prepared using the same amount of initiator (5:3 [Br]/[AGU]) but with a different monomer concentration (50:3 and 60:3 [monomer]/[AGU], respectively). CNC-METAC-2B was synthesized with 5:12 [Br]/[AGU]) and 60:3 [monomer]/[AGU] (Table S1). The colloidal suspensions of the unmodified and modified CNCs were prepared at 1 mg/mL in ultrapure water by vortexing for 15 sec followed by sonication (70 output) for 2 min. The sonicated suspensions were filtered using a 0.45 μm polytetrafluoroethylene (PTFE) filter and autoclaved at 121° C 15 psi for 30 min. The sterile CNCs suspensions were aliquoted and kept at −20 °C. For simplicity, the names of the modified cationic CNCs were abbreviated in the figures by omitting "CNC" from their denominations. For full name, chemical structures, and respective physicochemical characteristics, see Table S1 and Scheme 1.

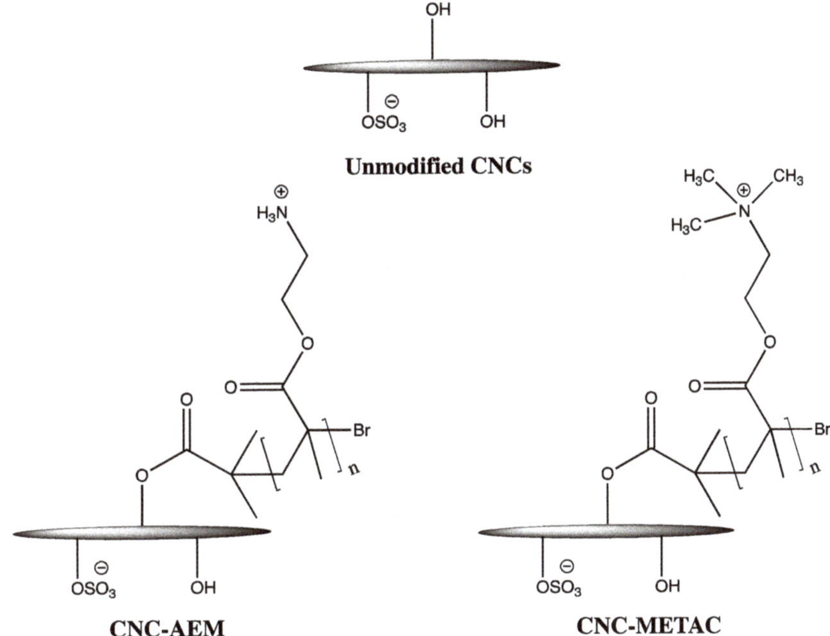

Scheme 1. Chemical representation of the cellulose-based nanomaterials used in this study: unmodified CNCs and engineered cationic CNCs (CNC-AEM and CNC-METAC).

2.2. *Human Blood Preparation*

Human blood containing citrate phosphate double dextrose Solution (CP2D) as an anticoagulant were extracted from Leukotrap blood filters (UVM Health Network-CVPH North Country Regional Blood Center) from healthy blood donors. To retrieve blood cells, the filter was slowly flushed once with a 50 mL syringe filled with air and approximately 15 mL was collected. The blood was diluted 1:10 in sterile calcium and magnesium-free phosphate buffered saline (PBS) to obtain approximately 1.5 mg/mL of total hemoglobin. In a 48-well plate, 225 μL of diluted blood was mixed with 25 μL of CNCs suspensions (final concentrations of 25 and 50 μg/mL), followed by incubation at 37 °C in a 5% CO_2-supplemented atmosphere for 1, 2, and 24 h before analysis.

2.2.1. Hemolysis Assay

The hemolysis assay was modified using the protocol from the National Cancer Institute (NCI) [23] and as previously published [24]. After the respective period of incubation, the percentage (%) of RBC hemolysis was analyzed by transferring 150 μL of the RBC/CNC mixture into a clean microtube and centrifuging it at 4000 RPM. The supernatant was then reacted with 1 mL of Drabkin solution (Ricca, Arlington, TX, USA) for 10 min RT in the dark, followed by spectrophotometric analysis at 540 nm. To determine the total hemoglobin in each experimental condition, 1% triton was added to 20 μL of the RBC/CNCs mixture for the complete release of hemoglobin prior to the addition to the 1 mL of Drabkin solution and the absorbance was measured at 540 nm using a microplate reader (Synergy H1 Hybrid Multi-Mode, BioTek/Agilent, Winoski, VT, USA). PBS (negative) ultrapure water, Triton 1%, and polyethylene glycol (PEG) 1 mg/mL (positive) were utilized as controls. The absorbance of the Drabkin solution at 540 nm was considered blank and subtracted from the absorbance obtained from all the samples. The percent of RBC lysis was calculated using the following equation.

$$\%\ \text{lysis} = \left[\frac{(\text{Abs540 nm of supernatants} - \text{Abs540 nm of blank})}{(\text{Abs540 nm of suspension} - \text{Abs540 nm of blank})\ x\ diluition\ factor}\right] \times 100$$

2.2.2. Red Blood Cell (RBC) Morphology and Aggregation

After the treatments, 10 μL of blood/CNCs mix were diluted in 100 μL of sterile calcium and magnesium-free PBS, and were promptly analyzed in a BD Accuri C6 Flow cytometer (BD Biosciences, Franklin Lakes, NJ, USA). At least 20,000 events were collected using a forward scatter channel (FSC) and sideward scatter channels (SCC), and were presented as histograms (FSC vs. count). The data were acquired and analyzed with BD Accuri software. PEG 1 mg/mL (positive control) was utilized for gating the aggregated cells.

For morphological changes and the aggregation of RBC, samples were diluted as described for flow cytometry analysis and observed in an Olympus CKX53 inverted microscope coupled with a DP22 Olympus camera. The Cell Sens (Olympus, Waltham, MA, USA) software was utilized to capture the images in bright field at 400x magnification.

2.3. *Cell Culture and Experimental Conditions*

Human umbilical vein endothelial cells (HUVEC, ATCC Manassas, VA, USA) were cultured in F-12K Medium (Kaighn's Modification of Ham's F-12 Medium, ATCC), supplemented with 10% fetal bovine serum (FBS, Gibco/Thermo Fisher Scientific, Waltham, MA, USA), 0.1 mg/mL heparin (MilliporeSigma, Burlington, MA, USA), and 0.3 mg/mL Endothelial Cell Growth Supplement (Corning, Corning, NY, USA). Cells were seeded at 1×10^5 cells/mL in a 96-well plate and cultured at 37 °C in a 5% CO_2-supplemented atmosphere for at least the overnight before the treatment with 10, 25, 50, and 100 μg/mL of CNCs for 24 or 48 h.

Cell Viability Assays

To assess the impact of CNCs on endothelial cell viability, the MTT assay (3-(4,5-dimethylthiazol-2-yl)-2,5-diphenyltetrazolium bromide (Sigma)) and Neutral Red assay (NR, Sigma) were utilized. These assays have different approaches to assess cell viability but both focus on organelle function. The MTT assay assesses the conversion of the water-soluble MTT (yellow) into a water-insoluble formazan (purple/blue), mainly by mitochondrial dehydrogenases [24]. The NR assay is based on the ability of viable cells to incorporate and bind the neutral red dye in the lysosomes [25]. In both assays, the intensity of the color is directly proportional to cell viability. After treatments, the medium from the HUVEC culture was removed and 100 μL of fresh culture medium containing 500 μg/mL of MTT or 50 μg/mL of NR was added to each well. The cells with the MTT or NR loading medium were incubated at 37 °C in a 5% CO_2-supplemented atmosphere. After 30 min, the respective loading medium was removed and the attached cells were gently washed once with PBS. To solubilize the formazan crystals, 100 μL/well of dimethyl sulfoxide (DMSO) was added and the absorbance was measured at 570 nm using a microplate reader (Synergy H1 Hybrid Multi-Mode, BioTek/Agilent). To extract NR dye from the lysosomes, 100 μL of acidified ethanol (1% glacial acetic acid, 50% ethanol) was added and the plate was placed on the plate shaker for ~10–15 min with protection from light. The absorbance at 540 nm and 690 nm was measured in the microplate reader within 60 min from adding the NR Desorb solution. For calculations, absorbance at 690 nm was subtracted from 540 nm. The non-treated cells (control) were considered 100% viable cells in both the MTT and NR assays. For statistical significance, both cell viability assays were repeated at least 3 times in triplicates.

2.4. Statistical Analysis

The data were statistically analyzed by using the two-way analysis of variance (ANOVA) test followed by a comparison test using GraphPad Prism 8.2 software. Multiple comparison tests and statistical significance are indicated in the legend of the respective figures.

3. Results and Discussion

Blood Compatibility Assay

First, we evaluated the interaction of unmodified CNCs and the respective cationic derivatives with human RBC, as these nanomaterials will be in contact with RBC when entering the circulatory system. Red blood cells (RBC) constitute almost half of blood volume [25] and therefore are important cells to determine the hemocompatibility of a nanomaterial. Initially, we assessed the capability of unmodified and modified CNCs to cause RBC lysis. Hemolysis refers to the damage of red blood cells leading to the release of intracellular content. A percent hemolysis less than 2 means that the nanoparticle is not hemolytic; 2–5% hemolysis means that the nanoparticle is slightly hemolytic; and >5% hemolysis means that the test sample is hemolytic [23]. Most of the CNCs demonstrated none or just slight hemolytic activity according to the definition above. In addition, we did not observe differences in hemolytic activity between the PBS (negative control) and any of the two concentrations of CNCs at a short period of exposure, namely 1 h (Figure S1) and 2 h (Figure 1A,B). As expected, triton 1% showed 100% hemolysis (data not shown) and PEG 1 mg/mL induced significant hemolysis with 2 and 24 h treatment, whereas ultrapure water only showed hemolytic activity at 24 h of incubation (Figure 1A, insert). At 24 h of exposure, however, 25 μg/mL of unmodified CNCs induced hemolysis at the threshold (5% or greater), notably significantly greater than the PBS (Figure 1A, black bars). The hemolytic effect of the unmodified CNCs on the RBC lysis could be attributed, at least in part, to the surface chemistry and reactivity between unmodified and modified CNCs. Unmodified CNCs displayed a negative surface charge (-34.8 ± 2.16 mV) and the derivatized CNCs showed a cationic surface charge ranging between $+31.8 \pm 2.89$ and $+45.0 \pm 1.44$ mV [20]. In addition to the charge differences, the reactivity of unmodified CNCs is also different from the derivatized counterparts. The unmodified CNCs contained greater amounts of hydroxyl and sulfate half-ester functional groups, and this abundance could lead to greater interactions with RBC membranes. For instance, hydroxyl moieties on silica NPs have been implicated in the hemolytic activity of silica as this functional group promotes interactions with cell membranes [26]. Additionally, the significant hemolytic effect of the unmodified CNCs occurred at a lower concentration, rather than higher. This apparent unexpected result could be due to the capability of unmodified CNCs to agglomerate at higher concentrations and therefore impact their interaction with cellular membranes. The degree of agglomeration of CNCs in cell culture media is among the several factors that can impact the results of biocompatibility assays [27].

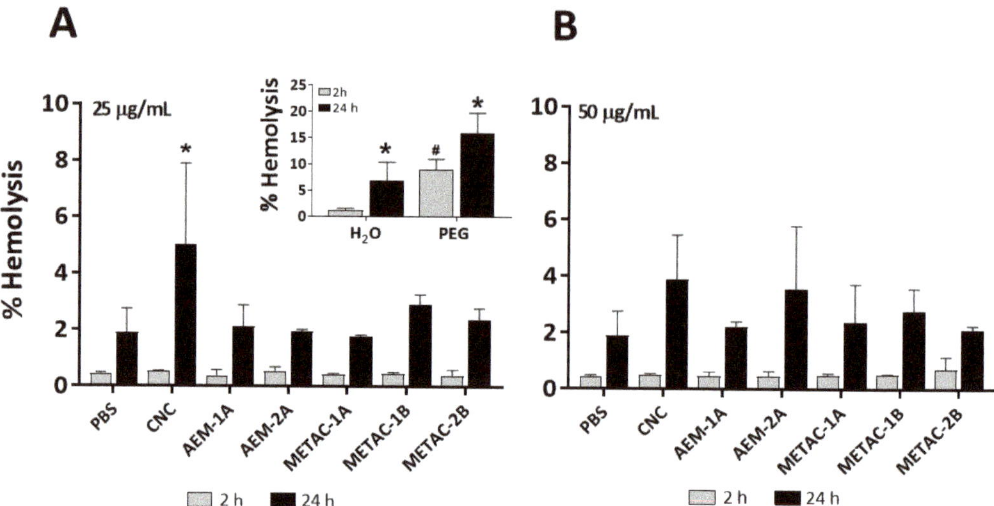

Figure 1. Hemolytic activity of unmodified and modified CNCs in human red blood cells (RBC). Percentage of hemolysis induced by 25 μg/mL (**A**) and 50 μg/mL (**B**) after 2 h (gray bars) and 24 h (black bars) of exposure in diluted human blood with respective concentrations of CNCs. PEG 1 mg/mL, ultra-pure H_2O, and Triton-X-100 1% (100% hemolysis, not shown) were used as positive controls (insert in **A**) and PBS as negative controls (**A**,**B**). * $p < 0.05$ vs. PBS 24 h; # $p < 0.05$ vs. PBS 2 h (Tukey's).

In parallel, we also investigated the possibility of unmodified CNCs and their cationic derivatives to cause RBC aggregation or changes in the RBC morphology using cell imaging and flow cytometry approaches. RBC aggregation and morphology together with RBC lysis are important parameters to assess the biocompatibility of nanomaterials [22]. None of the CNCs induced RBC aggregation at short (Figure 2A,B) or long exposure (Figure 3A,B), despite their cationic surface charges [20] and cationic polymers such as polyethyleneimine (PEI) that are well known to cause RBC aggregation [28]. The positive control, PEG 1 mg/mL, induced strong RBC aggregation in both short and long exposure (Figures 2 and 3C, respectively). The flow cytometry data confirmed what was observed in the cell imaging analysis. Typically, side scatter (SSC) signals are attributed to cellular internal structure and organelles, representing cellular granularity, and the forward scatter (FSC) signal is proportional to the diameter of the cell, representing the size of the cells [29]. None of the cationic CNCs or unmodified CNCs caused RBC aggregation at short or longer exposure (Figure 4A,B, respectively). PEG 1 mg/mL as a positive control was used to gate the "aggregation" behavior of the RBC. Although there was no major difference between RBC exposed to PBS (negative control) and cells exposed to unmodified and modified CNCs, we observed that the histogram profile had changed over time. At longer exposure (Figure 4B), the histogram displays two peaks for all the conditions, instead of one peak with 2 h of exposure (Figure 4A). RBC aggregation is expected to increase the FSC by shifting the histogram to the right, as observed in the positive control PEG (Figures 2 and 3C). The histogram of RBC exposed to 50 μg/mL of CNC-METAC-2B showed a mild increase in the number of events (counts) in the same region but not a right shift as expected for the increasing size of the cells which is an indication for RBC aggregation. This effect is more evident at 2 h (Figure 4A, right panel, green line) than at 24 h (Figure 4B, right panel, green line) of treatment. We did not observe differences in cellular granularity in all the conditions tested (Figure S2).

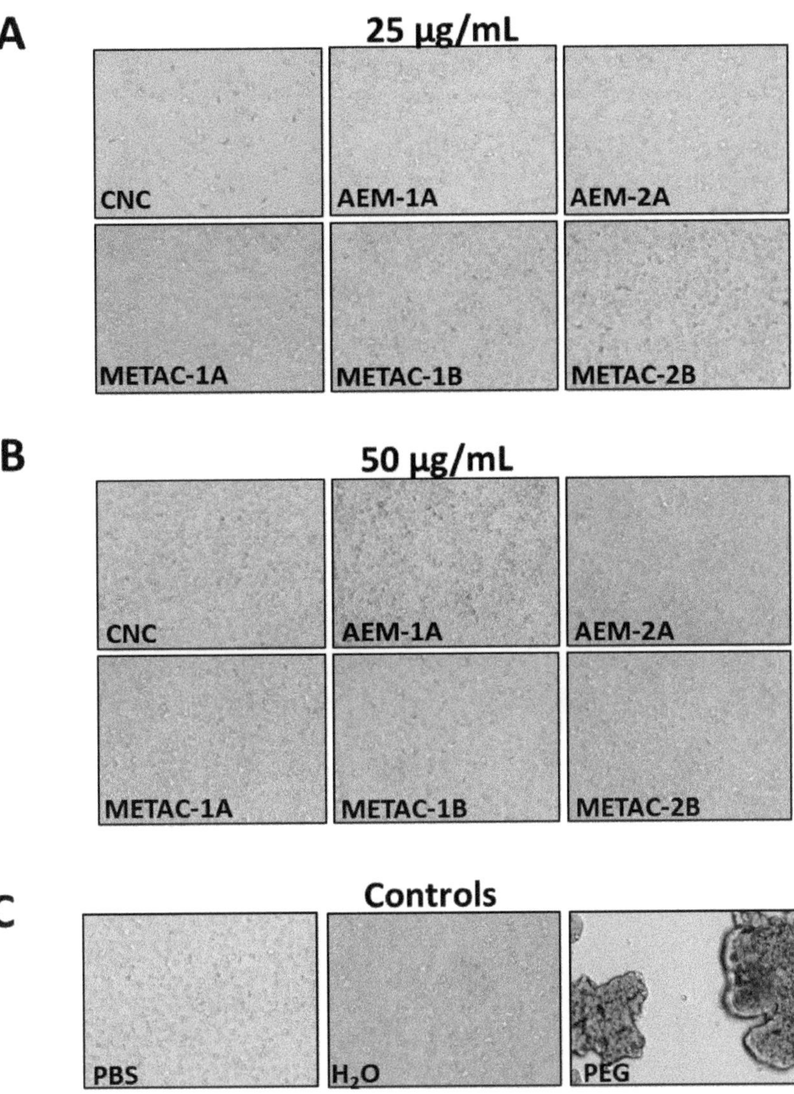

Figure 2. Effect of unmodified and modified CNCs on RBC morphology and aggregation. Diluted human blood was treated for 2 h with (**A**) 25 µg/mL or (**B**) 50 µg/mL of unmodified and modified CNCs, or with controls, (**C**) PBS, ultrapure H_2O, and PEG 1 mg/mL. The pictures were captured using a bright-field inverted microscope (400×).

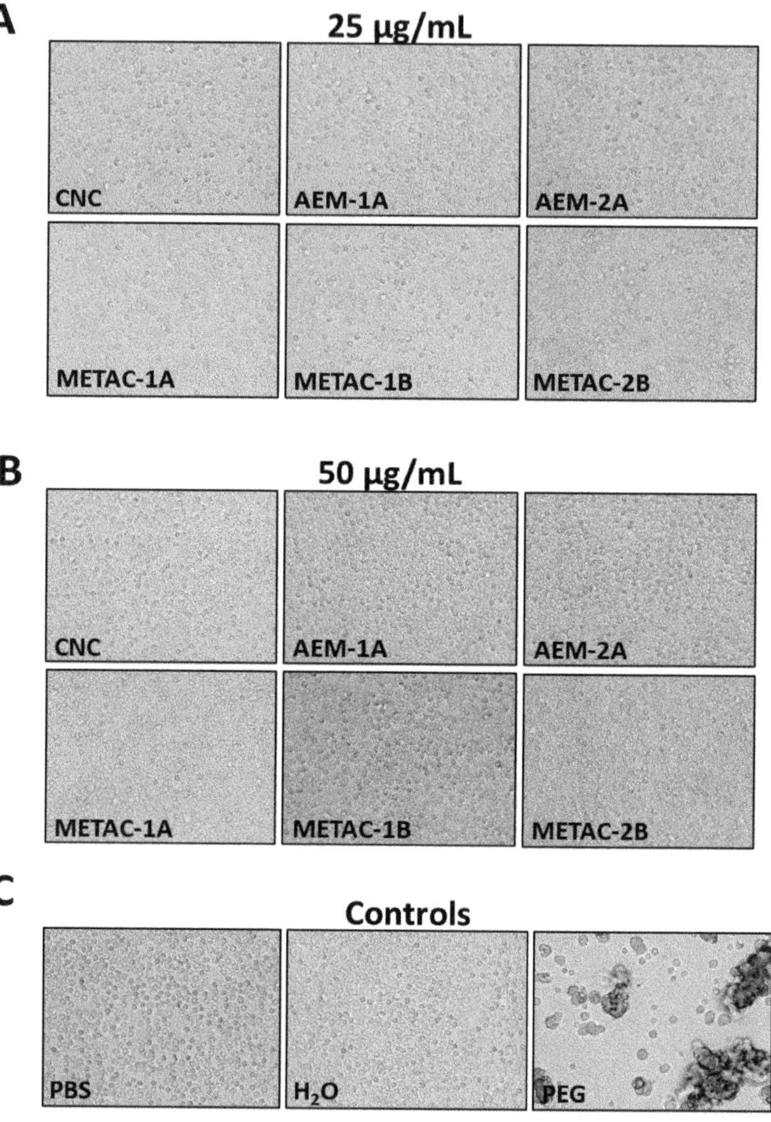

Figure 3. Effect of unmodified and modified CNCs on RBC morphology and aggregation. Diluted human blood was treated for 24 h with (**A**) 25 µg/mL or (**B**) 50 µg/mL of unmodified and modified CNCs, or with controls, (**C**) PBS, ultrapure H_2O, and PEG 1 mg/mL. The pictures were captured using a bright-field inverted microscope (400×).

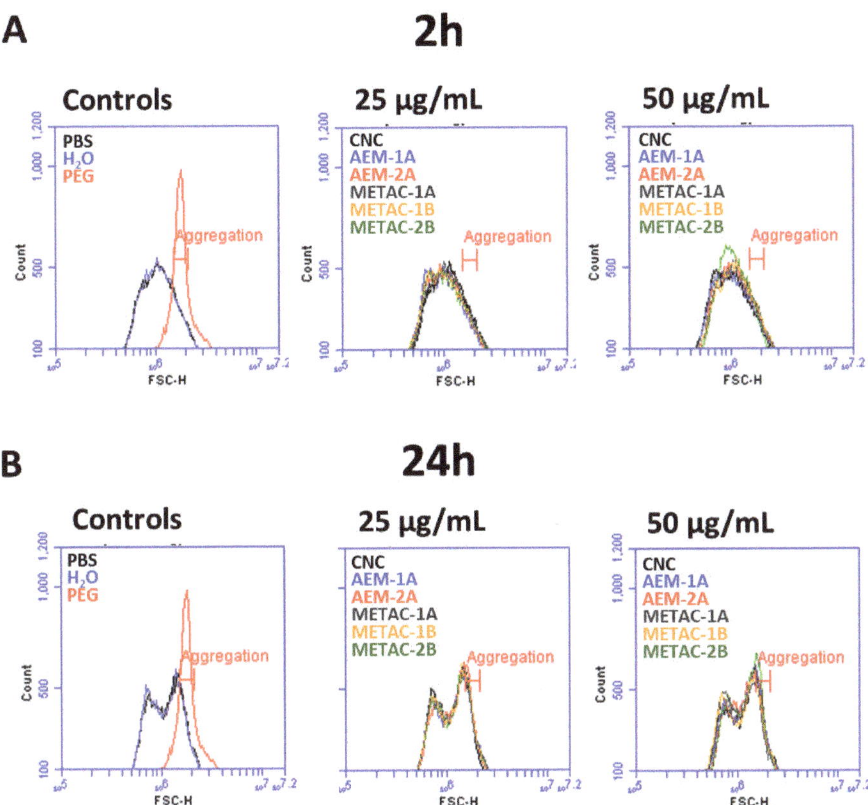

Figure 4. Effect of unmodified and modified CNCs on red blood cell (RBC) aggregation. Diluted human blood were treated for 2 h (**A**) or 24 h (**B**) with controls, PBS, ultrapure H_2O, and PEG 1 mg/mL (2 h only), or with 25 µg/mL or 50 µg/mL of unmodified and modified CNCs. After respective treatments, RBC aggregation was accessed by flow cytometry. The histograms displayed the number of events vs. the FSC-H (forward light scatter) high channel (size estimation). The aggregation of RBC was gated using PEG 1 mg/mL (red, 2 h) for both 2 and 24 h time points (**A**,**B**).

The surface chemistry and the morphology (size, shape, and state of aggregation) of nanoparticles that are in contact with the cells are important physicochemical aspects that drive the cytotoxicity of these nanomaterials [26]. Considering the growing interest in nanomaterials for biomedical applications, assessment of the toxicity in cell-based assays has become a crucial part of the characterization of nanomaterials [30]. In addition to the analysis of the impact of these cellulose-based nanomaterials on RBC and as part of the assessment of the biocompatibility of CNCs for potential biomedical applications, we chose to evaluate the effect of unmodified and engineered cationic CNCs on the cell viability of endothelial cells. Endothelial cells form a single cell layer that lines all blood vessels and regulates exchanges between the bloodstream and the surrounding tissues [31]. They are relevant in this context as these nanomaterials can reach systemic circulation and potentially be in contact with these cells; thus, they have been used to assess the biocompatibility of nanomaterials [28]. To evaluate the cytotoxicity of the cationic CNCs, we chose to perform MTT and NR assays because these low cost and reproducible assays are largely used for the screening of the cytotoxicity of compounds in general with potential biomedical applications. Most of the CNCs materials tested did not show a statistically significant decrease in the viability of HUVECs with 24 h of treatment, as demonstrated by MTT (Figure 5A) or NR (Figure 5B) assays. The exception is CNC-AEM-2A that at the highest concentration showed approximately a 20% decrease of cell viability using the NR assay (Figure 5B). The overall result is consistent with our previous work, in which we demonstrated that the same nanomaterials did not cause a decrease in the viability of human peripheral blood mononuclear cells (PBMCs) [20]. At 48 h of treatment, all nanomaterials, including unmodified CNCs, showed some degree of decrease in cell viability in both assays (Figure 5C,D). However, we did not observe a relevant dose–response relationship, which is the central concept in toxicology [32]. In general, this data is also consistent with the hemolytic effect of these compounds. The lack of a relevant dose-response decrease in cell viability is not a surprise as they also did not show relevant hemolytic activity. It has been suggested that hemolytic activity is generally associated with the cytotoxicity of nanoparticles [26].

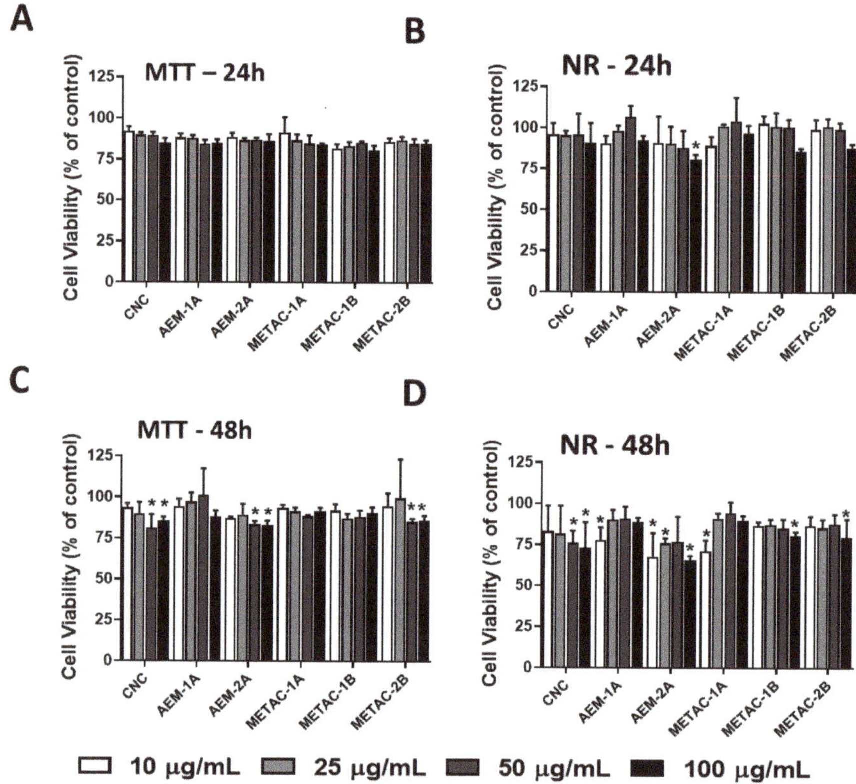

Figure 5. Effect of unmodified and modified CNCs on the cell viability of the HUVEC culture assessed by MTT (**A**,**C**) and NR (**B**,**D**) assays. After 24 h (**A**,**B**) or 48 h (**C**,**D**) of treatment, cell viability was determined spectrophotometrically using the respective assays. Representative of three independent experiments in triplicates. * $p < 0.01$ compared to the control, non-treated cells, (100% cell viability, Dunnett's).

4. Conclusions

In conclusion, we have assessed the vascular and blood compatibility of a series of engineered cationic CNCs. The interaction of both unmodified CNCs and cationic CNCs with human RBC concerning to RBC aggregation or changes in the RBC morphology was evaluated using cell imaging and flow cytometry approaches. We also assessed the effect of unmodified and engineered cationic modified CNCs on the cell viability of endothelial cells using MTT and NR assays. Overall, our results indicated that unmodified and engineered cationic CNCs are compatible with RBC as well as with endothelial cells. The findings of this study, together with the immunomodulatory properties of these cationic CNCs previously published, support the development of engineered biocompatible cationic CNCs for biomedical applications, in particular as vaccine nanoadjuvants.

Supplementary Materials: The following are available online at https://www.mdpi.com/article/10.3390/nano11082072/s1: Table S1: Composition of cationic CNCs as well as their respective zeta potential and apparent particle sizes; Figure S1: Hemolytic activity of unmodified and modified CNCs at 1 h of exposure; and Figure S2: Side scatter flow cytometer histograms of blood exposed for 2 and 24 h at different concentrations of unmodified and modified CNCs.

Author Contributions: Conceptualization, R.S. and K.C.; investigation, Y.I., B.T., C.W.S. (synthesis and characterization), A.B. and T.T. (flow cytometry and image analysis), H.N. and S.K. (hemolysis data analysis), and A.B. and K.C. (cell viability assays); data curation, Y.I., B.T. and K.C.; writing—review and editing, R.S. and K.C.; supervision, R.S. and K.C.; project administration, R.S. and K.C.; funding acquisition, R.S. and K.C. All authors have read and agreed to the published version of the manuscript.

Funding: This study is based upon work supported by the National Science Foundation under grant number 1703890.

Data Availability Statement: The data presented in this study are available in the main manuscript and in the Supplementary Materials.

Acknowledgments: The authors would like to thank InnoTech Alberta Inc., Edmonton, AB, Canada, for generously providing spray-dried sulfated CNCs for this study.

Conflicts of Interest: The authors declare no conflict of interest.

References

1. Jackson, J.K.; Letchford, K.; Wasserman, B.Z.; Ye, L.; Hamad, W.Y.; Burt, H.M. The use of nanocrystalline cellulose for the binding and controlled release of drugs. *Int. J. Nanomed.* **2011**, *6*, 321–330. [CrossRef]
2. Dong, S.; Roman, M. Fluorescently labeled cellulose nanocrystals for bioimaging applications. *J. Am. Chem. Soc.* **2007**, *129*, 13810–13811. [CrossRef]
3. Nielsen, L.J.; Eyley, S.; Thielemans, W.; Aylott, J.W. Dual fluorescent labelling of cellulose nanocrystals for pH sensing. *Chem. Commun.* **2010**, *46*, 8929–8931. [CrossRef]
4. Halib, N.; Ahmad, I. Nanocellulose: Insight into Health and Medical Applications. In *Handbook of Ecomaterials*; Martínez, L.M.T., Kharissova, O.V., Kharisov, B.I., Eds.; Springer International Publishing: Cham, Switzerland, 2019; pp. 1345–1363.
5. Kedzior, S.A.; Gabriel, V.A.; Dube, M.A.; Cranston, E.D. Nanocellulose in Emulsions and Heterogeneous Water-Based Polymer Systems: A Review. *Adv. Mater.* **2020**, *33*, e2002404. [CrossRef]
6. Habibi, Y.; Lucia, L.A.; Rojas, O.J. Cellulose nanocrystals: Chemistry, self-assembly, and applications. *Chem. Rev.* **2010**, *110*, 3479–3500. [CrossRef] [PubMed]
7. Moon, R.J.; Martini, A.; Nairn, J.; Simonsen, J.; Youngblood, J. Cellulose nanomaterials review: Structure, properties and nanocomposites. *Chem. Soc. Rev.* **2011**, *40*, 3941–3994. [CrossRef] [PubMed]
8. Klemm, D.; Kramer, F.; Moritz, S.; Lindstrom, T.; Ankerfors, M.; Gray, D.; Dorris, A. Nanocelluloses: A new family of nature-based materials. *Angew. Chem.* **2011**, *50*, 5438–5466. [CrossRef] [PubMed]
9. Vanderfleet, O.M.C.; Emily, D. Production routes to tailor the performance of cellulose nanocrystals. *Nat. Rev. Mater.* **2021**, *6*, 124–144. [CrossRef]
10. Sunasee, R. Nanocellulose: Preparation, Functionalization and Applications. In *Comprehensive Glycoscience*, 2nd ed.; Barchi, J.J., Ed.; Elsevier: Oxford, UK, 2021; pp. 506–537.
11. Sunasee, R.; Hemraz, U.D.; Ckless, K. Cellulose nanocrystals: A versatile nanoplatform for emerging biomedical applications. *Expert. Opin. Drug Deliv.* **2016**, *13*, 1243–1256. [CrossRef] [PubMed]
12. Huang, Y.P.; He, T.B.; Cuan, X.D.; Wang, X.J.; Hu, J.M.; Sheng, J. 1,4-beta-d-Glucomannan from Dendrobium officinale Activates NF-small ka, CyrillicB via TLR4 to Regulate the Immune Response. *Molecules* **2018**, *23*, 2658. [CrossRef] [PubMed]
13. Mold, M.; Shardlow, E.; Exley, C. Insight into the cellular fate and toxicity of aluminium adjuvants used in clinically approved human vaccinations. *Sci. Rep.* **2016**, *6*, 31578. [CrossRef]
14. Muñoz-Wolf, N.; McCluskey, S.; Lavelle, E.C. Chapter 2—The Role of Inflammasomes in Adjuvant-Driven Humoral and Cellular Immune Responses. In *Immunopotentiators in Modern Vaccines*, 2nd ed.; Schijns, V.E.J.C., O'Hagan, D.T., Eds.; Academic Press: Cambridge, MA, USA, 2017; pp. 23–42.
15. Gause, K.T.; Wheatley, A.K.; Cui, J.; Yan, Y.; Kent, S.J.; Caruso, F. Immunological Principles Guiding the Rational Design of Particles for Vaccine Delivery. *ACS Nano* **2017**, *11*, 54–68. [CrossRef] [PubMed]

16. Hogenesch, H. Mechanism of immunopotentiation and safety of aluminum adjuvants. *Front. Immunol.* **2012**, *3*, 406. [CrossRef] [PubMed]
17. Calabro, S.; Tritto, E.; Pezzotti, A.; Taccone, M.; Muzzi, A.; Bertholet, S.; De Gregorio, E.; O'Hagan, D.T.; Baudner, B.; Seubert, A. The adjuvant effect of MF59 is due to the oil-in-water emulsion formulation, none of the individual components induce a comparable adjuvant effect. *Vaccine* **2013**, *31*, 3363–3369. [CrossRef]
18. Sunasee, R.; Araoye, E.; Pyram, D.; Hemraz, U.D.; Boluk, Y.; Ckless, K. Cellulose nanocrystal cationic derivative induces NLRP3 inflammasome-dependent IL-1β secretion associated with mitochondrial ROS production. *Biochem. Biophys. Rep.* **2015**, *4*, 1–9. [CrossRef] [PubMed]
19. Guglielmo, A.; Sabra, A.; Elbery, M.; Cerveira, M.M.; Ghenov, F.; Sunasee, R.; Ckless, K. A mechanistic insight into curcumin modulation of the IL-1beta secretion and NLRP3 S-glutathionylation induced by needle-like cationic cellulose nanocrystals in myeloid cells. *Chem. Biol. Interact.* **2017**, *274*, 1–12. [CrossRef]
20. Imtiaz, Y.; Tuga, B.; Smith, C.W.; Rabideau, A.; Nguyen, L.; Liu, Y.; Hrapovic, S.; Ckless, K.; Sunasee, R. Synthesis and Cytotoxicity Studies of Wood-Based Cationic Cellulose Nanocrystals as Potential Immunomodulators. *Nanomaterials* **2020**, *10*, 1603. [CrossRef]
21. Neun, B.W.; Ilinskaya, A.N.; Dobrovolskaia, M.A. Updated Method for In Vitro Analysis of Nanoparticle Hemolytic Properties. *Methods Mol. Biol.* **2018**, *1682*, 91–102. [CrossRef]
22. Zhao, L.; Kaewprayoon, W.; Zhou, H.; Georgieva, R.; Baumler, H. RBC aggregation in dextran solutions can be measured by flow cytometry. *Clin. Hemorheol. Microcirc.* **2017**, *65*, 93–101. [CrossRef]
23. Neun, B.W.; Cedrone, E.; Dobrovolskaia, M.A. NCL Method ITA-1: Analysis of Hemolytic Properties of Nanoparticles. *NCIP Hub* **2020**. [CrossRef]
24. Ahmed, M.; Lai, B.F.; Kizhakkedathu, J.N.; Narain, R. Hyperbranched glycopolymers for blood biocompatibility. *Bioconjug. Chem.* **2012**, *23*, 1050–1058. [CrossRef] [PubMed]
25. Narain, R.; Wang, Y.; Ahmed, M.; Lai, B.F.; Kizhakkedathu, J.N. Blood Components Interactions to Ionic and Nonionic Glyconanogels. *Biomacromolecules* **2015**, *16*, 2990–2997. [CrossRef]
26. Zhang, H.; Dunphy, D.R.; Jiang, X.; Meng, H.; Sun, B.; Tarn, D.; Xue, M.; Wang, X.; Lin, S.; Ji, Z.; et al. Processing pathway dependence of amorphous silica nanoparticle toxicity: Colloidal vs. pyrolytic. *J. Am. Chem. Soc.* **2012**, *134*, 15790–15804. [CrossRef] [PubMed]
27. Colic, M.; Tomic, S.; Bekic, M. Immunological aspects of nanocellulose. *Immunol. Lett.* **2020**, *222*, 80–89. [CrossRef] [PubMed]
28. Xia, Y.; Cheng, C.; Wang, R.; Qin, H.; Zhang, Y.; Ma, L.; Tan, H.; Gu, Z.; Zhao, C. Surface-engineered nanogel assemblies with integrated blood compatibility, cell proliferation and antibacterial property: Towards multifunctional biomedical membranes. *Polym. Chem.* **2014**, *5*, 5906–5919. [CrossRef]
29. Givan, A.L. Principles of flow cytometry: An overview. *Methods Cell Biol.* **2001**, *63*, 19–50. [CrossRef]
30. Nel, A.; Xia, T.; Meng, H.; Wang, X.; Lin, S.; Ji, Z.; Zhang, H. Nanomaterial Toxicity Testing in the 21st Century: Use of a Predictive Toxicological Approach and High-Throughput Screening. *Acc. Chem. Res.* **2013**, *46*, 604–621. [CrossRef]
31. Alberts, B.; Johnson, A.; Lewis, J.; Raff, M.; Roberts, K.; Walter, P. Blood Vessels and Endothelial Cells. In *Molecular Biology of the Cell*; Garland Science: New York, NY, USA, 2002.
32. Schulte-Hermann, R.; Grasl-Kraupp, B.; Bursch, W. Dose-response and threshold effects in cytotoxicity and apoptosis. *Mutat. Res.* **2000**, *464*, 13–18. [CrossRef]

nanomaterials

MDPI

Article

Tracking Bacterial Nanocellulose in Animal Tissues by Fluorescence Microscopy

Renato Mota [1,2], Ana Cristina Rodrigues [1,2], Ricardo Silva-Carvalho [1,2], Lígia Costa [1,2], Daniela Martins [1,2], Paula Sampaio [3,4], Fernando Dourado [1,2] and Miguel Gama [1,2,*]

[1] CEB—Centre of Biological Engineering, University of Minho, Campus Gualtar, 4710-057 Braga, Portugal; renatovelosomota@gmail.com (R.M.); anacris38599@gmail.com (A.C.R.); remanuelcarvalho@gmail.com (R.S.-C.); lfr.costa92@gmail.com (L.C.); dsr.martins92@gmail.com (D.M.); fdourado@deb.uminho.pt (F.D.)

[2] LABBELS—Associate Laboratory, 4710-057 Braga, Portugal

[3] Instituto de Investigação e Inovação em Saúde, Universidade do Porto, 4200-135 Porto, Portugal; sampaio@ibmc.up.pt

[4] IBMC—Instituto de Biologia Celular e Molecular, 4200-135 Porto, Portugal

* Correspondence: fmgama@deb.uminho.pt; Tel.: +351-253-604-400

Abstract: The potential of nanomaterials in food technology is nowadays well-established. However, their commercial use requires a careful risk assessment, in particular concerning the fate of nanomaterials in the human body. Bacterial nanocellulose (BNC), a nanofibrillar polysaccharide, has been used as a food product for many years in Asia. However, given its nano-character, several toxicological studies must be performed, according to the European Food Safety Agency's guidance. Those should especially answer the question of whether nanoparticulate cellulose is absorbed in the gastrointestinal tract. This raises the need to develop a screening technique capable of detecting isolated nanosized particles in biological tissues. Herein, the potential of a cellulose-binding module fused to a green fluorescent protein (GFP–CBM) to detect single bacterial cellulose nanocrystals (BCNC) obtained by acid hydrolysis was assessed. Adsorption studies were performed to characterize the interaction of GFP–CBM with BNC and BCNC. Correlative electron light microscopy was used to demonstrate that isolated BCNC may be detected by fluorescence microscopy. The uptake of BCNC by macrophages was also assessed. Finally, an exploratory 21-day repeated-dose study was performed, wherein Wistar rats were fed daily with BNC. The presence of BNC or BCNC throughout the GIT was observed only in the intestinal lumen, suggesting that cellulose particles were not absorbed. While a more comprehensive toxicological study is necessary, these results strengthen the idea that BNC can be considered a safe food additive.

Keywords: food additive; bacterial nanocellulose; bacterial cellulose nanocrystals; cellulose binding module; fluorescence microscopy; gastrointestinal tract; absorption

Citation: Mota, R.; Rodrigues, A.C.; Silva-Carvalho, R.; Costa, L.; Martins, D.; Sampaio, P.; Dourado, F.; Gama, M. Tracking Bacterial Nanocellulose in Animal Tissues by Fluorescence Microscopy. *Nanomaterials* **2022**, *12*, 2605. https://doi.org/10.3390/nano12152605

Academic Editors: Rajesh Sunasee and Karina Ckless

Received: 28 June 2022
Accepted: 26 July 2022
Published: 28 July 2022

1. Introduction

The 21st century's environmental and economic challenges, in particular related to sustainability and safety, have been driving the preference for the use of green, renewable and recyclable raw materials for the production of high-value-added products but with lower environmental impact [1]. Nanocellulose (NC), a nano-scaled cellulosic material (with at least one dimension <100 nm), from plant or bacterial sources [1–4], is at the forefront of such promising biopolymers [2]. Due to its large specific area, negligible toxicity, low density, biodegradability and biocompatibility, NC has increasingly been explored in several fields such as textiles, pharmaceuticals, cosmetics, medicine, food and packaging [4–6]. In 2020, the worldwide market for NC has been estimated at USD 297 million and is expected to increase by 2025 to as much as USD 783 million [7].

As food additives, micron and colloidal plant cellulose and their derivatives are approved for human consumption by several regulatory authorities, such as the FDA (Food and Drug Administration, Silver Spring, MD, USA) and the EC (European Commission); they are regularly applied in the food industry to regulate its texture, stability, rheology and organoleptic features [8,9]. In the 1980s, plant nanocellulose was first suggested as a human food additive with enormous potential, as its high aspect ratio could confer new physicochemical properties and behaviors to foods [10]. Ever since, several other studies have pointed out the potential of NC to further improve food quality and appeal, as well as their capacity to modulate digestion and nutrient absorption [11]. But, because of their nanoscalar nature, different properties and interactions with biological systems may arise, as compared to the micro-sized counterparts [12,13], raising several concerns throughout the various stages of NC life cycle. This is especially relevant, as the biological effects of nanocelluloses depend on their chemical nature, size, shape, aggregation properties, degree of branching, specific surface properties, among others. These intrinsic properties of NC, affecting their interactions with cells and living organisms, are still poorly understood [14–16]. For instance, nanotoxicity derives mainly from the small size and large surface area of engineered nanomaterials, which may enable their translocation to different organs by getting absorbed into the blood through the intestinal lumen [17–19]. Several publications have reported the toxicity of ingested inorganic nanomaterials such Ag or ZnO nanoparticles in animal models [20–22]. Regarding plant or bacterial celluloses' absorption in the gastrointestinal tract (GIT), few and contradictory studies have been published [23–28]. Reports have shown that bacterial [23] and plant cellulose [8] are slowly and limitedly degraded in rats' large intestine, yielding metabolites that are partially absorbed by the colon and/or microflora, in both cases used mainly as an energy source. These cellulose degradation products, absorbed in the intestine, were detected in urine and in exhaled CO_2. Studies with germ-free rats (no intestinal microflora) were also carried out, wherein the total excretion of cellulose was observed, thereby concluding that there is no absorption of undegraded cellulose but only of its degradation products [23]. Thus, a thorough environmental and risk assessment (the fate and toxicity of NC in the human body) is of extreme importance when considering NC production (occupational exposure), commercialization and use (human consumption).

On the other hand, the lack of appropriate detection and characterization techniques and the absence of reproducible and validated methods for toxicological studies have been identified as major bottlenecks in the evaluation of the safety of nanomaterials [12,13,29–32]. For instance, the metabolization of cellulose by the colon microflora makes it very difficult to distinguish the celluloses' fibers from their degradation products, not allowing the clarification of whether the fibers are absorbed in the GIT. BNC is a nanofibrillar exopolysaccharide produced by acetic-acid-bacteria, such as the ones from the genera *Komagataeibacter*. Although chemically identical to plant cellulose, BNC is chemically pure and has higher tensile strength, water-holding capacity and crystallinity than cellulose from plant sources [4]. Furthermore, while plant cellulose comminution to the nanoscale requires mechanical, chemical or enzymatic processes, BNC is naturally nano-sized [4,9]. In food applications, BNC is marketed mostly in Asia as "nata de coco" [4,9,33], but it has been attracting the attention of many industries worldwide, given its unique features. The available toxicological data on BNC has been extensively reviewed by Dourado et al. (2016), wherein (i) the absence of genotoxicity, carcinogenicity, pyrogenicity or developmental or reproductive toxicity and (ii) the long history of its consumption (without any reported cases of toxicity) were exposed [33]. Like plant nanocellulose, the potential hazards of ingested BNC, derived from its nanoscale nature, are insufficiently characterized, hindering its entry into the Western food market.

Fluorescence-based detection is the most common method used in biosensing, due to its relatively low-cost and high sensitivity, specificity and simplicity. On the other hand, electron microscopy is the technique of choice when a high resolution is required [34–37]. To fill the gap between light and electron microscopies, correlative light and electron microscopy

(CLEM) strategies have been developed, allowing the establishment of a correlation between a particular ultrastructural feature (by SEM or TEM) and a fluorescence signal [34–37]. This approach could be used to determine the sensitivity of fluorescence microscopy (FM) and its potential to detect cellulose nanostructures in biological tissues, namely in the GIT, in particular BNC. Thus, the present study was designed to evaluate whether fluorescence analysis is sensitive enough to detect isolated bacterial cellulose nanocrystals (BCNC) using CLEM. As such, we prepared BCNC by acid hydrolysis [38–40] to simulate an (unlikely) scenario of the extreme digestion of BNC in the GIT. The BCNC were stained with a fluorescently labeled cellulose-binding domain [41], and, using CLEM, we attempted to determine whether isolated nanocrystals are detected by FM. After such validation, fluorescence microscopy was used in the analysis of the histological slides of different tissues from Wistar rats, fed with BNC, to determine whether the fibers (or their degraded versions) could be detected, establishing a platform for more comprehensive studies.

2. Materials and Methods

2.1. Materials

BNC membranes were obtained from HTK Food Co., Ltd. (Ho Chi Minh City, Vietnam). Chemicals used for the reaction of the bacterial cellulose nanocrystals, namely H_2SO_4 and HCl, were supplied by Thermo Fisher Scientific (Waltham, MA, USA).

2.2. Methods

2.2.1. Preparation of Bacterial Nanocellulose

The as-received BNC membranes were washed with distilled water (dH_2O) until the pH became that of the dH_2O, to remove any soluble chemicals. Then, the membranes were cut into small pieces and ground using a high-speed blender (Moulinex Ultrabend1 500 W, Écully, France), at 24,000 rpm for 5 min, to obtain a pulp, which was then filtered. The BNC cake was further concentrated by centrifugation at 11,000 rpm (Centrifuge 5430 R, Eppendorf, Hamburg, Germany) for 20 min at room temperature (RT). The solid fraction of the centrifugate was adjusted to 10% (*m/v*) with dH_2O and stored at 4 °C in a glass container until use.

2.2.2. Production of Bacterial Cellulose Nanocrystals

The BNC centrifugate was subjected to acid hydrolysis with a solution of H_2SO_4/HCl (34 and 24% *m/m*, respectively), as described elsewhere [39]. For each batch of BCNC production, 10 g of BNC (at 10% solids) and 1000 mL of acid solution were mixed at 45 °C with magnetic stirring (500 rpm) with a H03D mini-stirrer (lbx Instruments, Lonay, Switzerland) for 75 min. The hydrolysis reaction was then stopped by diluting the reaction 15-fold with cold dH_2O. The suspension was ultracentrifugated at 11,000 rpm (Heraeus Multifuge X3R, Thermo Fisher Scientific, Waltham, MA, USA) for 15 min at RT to precipitate the BCNC, which were then washed with dH_2O under several cycles of ultracentrifugation (11,000 rpm for 20 min) to remove excess acid. This procedure was repeated until the pH was in the range 5–7. The resulting suspension was sonicated for 3 min at 500 W (Sonics & Materials, Newtown, CT, USA), and the final concentration was adjusted to ~1% (*m/v*).

2.3. Characterization of the BCNC

2.3.1. Zeta Potential

The stability of the BCNC aqueous suspension was assessed by measuring the particles' surface zeta potential. For that, BCNC suspensions of 0.1% (*m/v*) in dH_2O were prepared and the zeta potential measured using a Zetasizer Nano ZS (Malvern Instruments, Malvern, UK). Data were calculated considering the viscosity of water at 25 °C of 0.893×10^{-3} Pa.s^{-1}. Three measurements of each suspension were performed. Throughout the work, six batches of BCNC were made. The zeta potential result displayed is the average of all measurements.

2.3.2. Fourier-Transform Infrared Spectroscopy (FTIR)

FTIR spectroscopic analysis of BNC and BCNC was carried out in a Bruker FTIR spectrometer ALPHA II (Bruker Corporation, Billerica, MA, USA) in transmission mode, operating at a resolution of 4 cm^{-1}. BNC and BCNC were both frozen and processed in a freeze-drier (Coolsafe 100-9 Pro, Labogene, Allerød, Denmark). Two milligrams of the samples were then mixed with 200 mg of dry potassium bromide (Thermo Fisher Scientific, Waltham, MA, USA) to obtain a solid pellet that was scanned three times to check the authenticity of data. The spectra were taken between 4000 and 700 cm^{-1} by averaging 24 scans for each spectrum.

2.3.3. Transmission Electron Microscopy (TEM)

The morphology of the BCNC was assessed by TEM imaging. Briefly, 5 μL of BCNC aqueous suspension at 0.01% (*m/v*) was applied on the grid (FCF400-Cu, Electron Microscopy Sciences (EMS), Hatfield, UK) and allowed to settle for 2 min at RT. The sample was then blotted off, and 5 μL of uranyl acetate (UA) (Sigma-Aldrich, St. Louis, MO, USA) for negative staining was directly applied. The excess solution was again blotted off and replaced by another 5 μL of UA. Each time, the UA was allowed to incubate for 30 s before it was removed. The sample was observed with a JEOL 2100 plus TEM device (JEOL, Tokyo, Japan), operated at 80 kV accelerating voltage. Several images were taken, considering areas far from each other and trying to represent the whole grid's surface. The length (L) and width (W) of the BCNC were determined from at least 150 measurements by image analysis, using ImageJ software (Version 1.51j8, Bethesda, MD, USA) [42].

2.4. GFP-CBM3A Adsorption onto BNC and BCNC

Knudsen et al. (2015) developed a method for the specific and sensitive detection of cellulose fibers, including nanofibrillar cellulose, in biological tissues (both in cryopreserved and paraffin-embedded tissues), using a biotinylated carbohydrate-binding module (CBM) of the β-1,4-glycanase from the bacterium *Cellulomonas fimi* [41]. Based on their work, we evaluated the efficiency of a GFP-fused CBM derived from *Clostridium cellulolyticum* (GFP–CBM3A, NZytech, Lisbon, Portugal) to bind both to BNC and BCNC. The recombinant GFP–CBM3A (henceforward designated as GFP–CBM, for the sake of simplicity) was purified from *Escherichia coli* and contains a family 3A carbohydrate-binding module (CBM3A) and an N-terminal green fluorescent protein (GFP). CBM from the type 3A family binds specifically to crystalline forms of cellulose [41].

For these experiments, a HORIBA scientific spectrofluorometer (Kyoto, Japan) was used, operating at emission and excitation wavelengths of 510 and 475 nm, respectively. A calibration curve using aqueous solutions of GFP–CBM at 0.005, 0.010, 0.0150, 0.020 and 0.025 mg/mL was first obtained.

Next, in eppendorf tubes, different solutions of GFP–CBM, with concentrations of 0.500, 0.250, 0.175, 0.100, 0.050, 0.025, 0.0125, 0.005 and 0.000 mg/mL in the final mixture, were each incubated with a fixed mass of 0.25 mg (dry basis) of BNC or BCNC in 0.40 mL of dH_2O for 2 h (sufficient time to reach equilibrium, as observed in exploratory assays) at RT. Then, the dispersions were centrifuged (Centrifuge 5430 R, Eppendorf, Hamburg, Germany) for 10 min at 8000× *g*. The GFP–CBM in the collected supernatant was quantified through spectrofluorometry, as described above.

A non-linear regression analysis was used to calculate the parameters of the Langmuir adsorption isotherm [43,44]:

$$\mathrm{GFP-CBM_{Bound}} = \frac{\mathrm{GFP-CBM_{Max}\ .\ K_a\ .GFP-CBM_{Free}}}{1+\ \mathrm{K_a\ .GFP-CBM_{Free}}} \quad (1)$$

where GFP–CBM_{Bound} is the amount of adsorbed protein per unit mass of cellulose (mg/mg), GFP–CBM_{Free} is the protein concentration (mg/mL) in the liquid phase at the adsorption equilibrium, GFP–CBM_{Max} is the maximum amount of adsorbed protein per unit mass of cellulose (mg/mg), and K_a is the Langmuir constant (mL/mg). Three independent assays were performed, each of them in triplicate. The non-linear regression and parameters (K_a and GFP–CBM_{Max}) were calculated using OriginPro 2018 software (Version 9.5.1.195, Northampton, MA, USA) [45].

Finally, the binding of the GFP–CBM was also assayed qualitatively by the observation of the GFP–CBM-bound celluloses by FM on an Olympus BX51 (Olympus, Tokyo, Japan) with a 60×objective and a FITC filter set.

2.5. *Correlative Light Electron Microscopy (CLEM)*

The feasibility of FM to detect individual nanometric particles was evaluated by CLEM. To this end, reference grids (50/B D300F1FC-50CU, EMS) with adsorbed BCNC were observed using widefield fluorescence microscopy and then by TEM to detect co-localized single BNC crystals. Briefly, suspensions of BCNC were diluted to 0.001 mg/mL and labeled with GFP–CBM at a final concentration of 0.05 mg/mL for 30 min at RT. Next, 5 µL of GF—CBM:BCNC suspension was applied to the grid and allowed to settle for 2 min. Then, the sample was blotted off and sequentially observed on (i) a FM Nikon ECLIPSE Ti (Nikon, Tokyo, Japan) with a 10x/0.45 or 40x/0.95 PlanApo objectives, FITC filter set and an IRIS 9 camera (Photometrics, Huntington Beach, CA, USA) under the control of NIS-Elements software (Version 4.5, Nikon, Tokyo, Japan) and then on (ii) a JEOL JEM 1400 Transmission Electron Microscope (JEOL, Tokyo, Japan). Images were digitally recorded using a CCD digital camera 1100W (Orius, Tokyo, Japan).

Some technical issues regarding sample throughput between FM and TEM observations arose, which were overcome by using gelatin from edible grade (Results and Discussion, Section 3.3). The effect of the gelatin matrix on the visualization of BCNC was first assessed in both FM and TEM. After optimization, 0.002 mg/mL BCNC labeled with GFP–CBM (0.05 mg/mL) (prepared as described above) were mixed with a solution of 2% gelatin (1:1 final mass ratio) for 15 min at 45 °C. After cooling to RT, the suspension was (i) placed on microscope slides (VWR, Radnor, PA, USA) for FM evaluation or (ii) applied to the grid (FCF200-NI-TA, EMS) and allowed to settle for 2 min before analysis. The length (L) and width (W) of the BCNC were determined using ImageJ (Version 1.51j8, Bethesda, MD, USA). Samples were processed for CLEM following the aforementioned methodology.

2.6. *In Vitro Assays*

2.6.1. Cell Line and Cell Culture

Mouse fibroblast L929 cell line was purchased from American Type Culture Collection (ATCC, Manassas, VA, USA). The cell line was maintained in Dulbecco's modified essential medium (DMEM) (Sigma-Aldrich, St. Louis, MO, USA) supplemented with 10% (*v/v*) iFBS (Sigma-Aldrich, St. Louis, MO, USA) and 1% (*v/v*) penicillin (10,000 U/mL, Sigma-Aldrich, St. Louis, MO, USA)–streptomycin (10,000 µg/mL, Invitrogen, Waltham, MA, USA) (complete DMEM medium) at 37 °C and 5% CO_2.

2.6.2. Bone-Marrow-Derived Macrophage (BMMΦ) Differentiation

Mouse bone marrow cells were differentiated in vitro as described elsewhere [46]. C57BL/6 mice were anesthetized using isoflurane and euthanized by cervical dislocation. Femurs and tibias were removed and cleaned in aseptic conditions. Bones were disconnected by the articulations and then flushed using complete DMEM medium. The obtained cell suspension of bone marrow cells was centrifuged (300× *g*, 10 min) at RT and the pellet resuspended in 15 mL of complete DMEM medium supplemented with 20% (*v/v*) L929-cell conditioned medium (LCCM) as a source of macrophage colony-stimulating factor (M-CSF). Prior to this, LCCM was prepared as follows: L929 cells (at a initial density of 5 × 10^3 cells/mL) were grown for 8 days at 37 °C in a 5% CO_2 atmosphere. Cells were then centrifuged (300× *g*, 10 min), the supernatant collected, filtered (0.2 µm filter) and finally stored at −20 °C before use. The bone marrow cell suspension was cultured overnight at 37 °C in a 5% CO_2 atmosphere. The non-adherent cells were collected with warm complete DMEM medium supplemented with 20% (*v/v*) LCCM, plated in a 96-well plate and incubated once again at 37 °C in a 5% CO_2 atmosphere. On the 4th and 7th days, half of the medium was renewed, and on the 10th day, cells became fully differentiated into macrophages.

2.6.3. Cytotoxicity Testing

The effect of BCNC on the viability of BMMΦ primary cells and in L929 cells was studied using a standard resazurin assay [47]. Briefly, a monolayer of BMMΦ (at a density of 3×10^4 cells/well) and L929 (at a density of 1×10^4 cells/well) were each incubated for 24 h (37 °C in a 5% CO_2 atmosphere) with increasing concentrations of BCNC (0.00, 0.001, 0.005, 0.01, 0.02, 0.04, 0.10, 0.20 and 1.00 mg/mL), previously sterilized by UV radiation for 15 min at RT. A higher BCNC concentration (2.0 mg/mL) was also used for cytotoxicity evaluation in BMMΦ cells. After incubation, 10% (*v/v*) of a 2.5 mM resazurin solution (Sigma-Aldrich) was added to each well, and the cells were incubated at 37 °C in a 5% CO_2 atmosphere for another 4 h. A well containing only LCCM-supplemented DMEM medium was used as the blank sample. Fluorescence was measured (λex 560/λem 590 nm) in a SpectraMAX GeminiXS microplate reader (Molecular Devices LLC, San Jose, CA, USA). Results were expressed as the mean percentage ± SD of viable cells relative to a control without BCNC (considered as 100% viability). Three independent assays were performed, each in triplicate.

2.6.4. Uptake of BCNC by BMMΦ Primary Cells

The uptake of BCNC by macrophages was evaluated either by FM and TEM. For that, 1 mL of BMMΦ cells (at a density of 1×10^6 cells/well) was incubated in 24-well plates (at 37 °C in a 5% CO_2 atmosphere) with a suspension of previously sterilized nanocrystals at a concentration of 0.001 mg/mL for (i) 2, 4, 6 and 24 h for fluorescence analysis and for (ii) 4 h and 24 h for TEM analysis.

For FM, inside each well was a chemically modified glass coverslip (ECN 631-1578, VWR) that allowed the macrophages to grow and adhere to its surface. After each incubation period, the medium was removed, and the cells were rinsed in fresh medium and fixed for 15 min in 4% paraformaldehyde (PFA) (Sigma-Aldrich, St. Louis, MO, USA). Subsequently, cells were rinsed 3 times with PBS (Sigma-Aldrich, St. Louis, MO, USA), permeabilized with 0.1% Triton X-100 (PanReac, Barcelona, Spain) at RT for 15 min and rinsed again with PBS. Nuclei were stained with 1 μg/mL DAPI (Frilabo, Porto, Portugal) at RT for 30 min. Cells were rinsed again with PBS, and then the actin-cytoskeleton was stained with 1 μg/mL phalloidin-TRITC (pha-red) (Sigma-Aldrich, St. Louis, MO, USA) for 30 min at RT. Cells were rinsed with PBS and incubated with GFP–CBM (0.05 mg/mL) for BCNC detection. After 3 final rinses in PBS, the coverslips were removed from the wells using tweezers and mounted for FM analysis. As control, macrophages without BCNC incubation were used. FM images of the stained macrophages were acquired using an Olympus BX61Confocal Scanning Laser Microscope (Model FluoView 1000, Olympus, Tokyo, Japan) using the following combination of excitation/emission detection-range wavelengths: 405/430–470, 488/505–540, and 559/575–675, for the visualization of the cells' nuclei, GFP–CBM:BCNC and cells' cytoskeleton, respectively. Images were acquired with the software FV10-Ver4.1.1.5 (Olympus, Tokyo, Japan).

For electron microscopy analysis, the cells grew and adhered directly to the bottom of the well plate. After the incubation period, trypsin was used to detach the cells, which were centrifuged for 10 min at 800× *g*. The resulting pellet was fixed in a solution of 2.5% glutaraldehyde (#16316; EMS) with 2% formaldehyde (#15713; EMS) in 0.1 M sodium cacodylate buffer (pH 7.4) for 1 day and post fixed in 1% osmium tetroxide (#19190; Electron Microscopy Sciences, Hatfield, UK) and 1.5% potassium ferrocyanide (Sigma-Aldrich, St. Louis, MO, USA) diluted in 0.1 M sodium cacodylate buffer (Sigma-Aldrich, St. Louis, MO, USA). After centrifugation, the pellet was washed in dH_2O and then stained with aqueous 1% UA solution overnight, dehydrated and embedded in Embed-812 resin (#14120; EMS). Ultra-thin sections (50 nm thickness) were cut on an RMC Ultramicrotome (RMC Boeckeler, Hamburg, Germany) using Diatome diamond knifes, mounted on mesh copper grids (EMS) and stained with UA substitute (#11000; EMS) and lead citrate (#11300; EMS) for 5 min each. As control, macrophages without BCNC incubation were used. Samples were viewed on the JEOL 1400 (Tokyo, Japan), and images were digitally recorded using a CCD digital camera (Orius 1100W, Tokyo, Japan).

2.7. Cellulose Tracking in Animal Tissues

2.7.1. Animals, Housing and Feeding Conditions

The study was performed at I3S. The experimental procedures followed the EU Directive 2010/63/EU and National Decreto-Lei 113/2013 legislation for animal experimentation and welfare. The rats' housing, handling and experimentation were accredited by the Portuguese National Authority for Animal Health, Direção-Geral de Alimentação e Veterinária (DGAV) (approval n°012910/2020-08-07).

This pilot study aimed at establishing the methodological grounds—namely concerning the detection of cellulose fibers by a histological analysis of the tissues collected from the rats—for a larger and more comprehensive study.

Eight-week-old Wistar Han IGS Rats (Crl:WI(Han)), four male and four female, were used in this study. The rats were obtained from Charles River (Barcelona, Spain) and bred at a i3S animal facility. On arrival, the animals were examined for signs of health, followed by a one-week adaptation period. They were kept under controlled environmental conditions for one week before starting the experiment. Each rat was uniquely numbered with a color marker in their tail and placed in polycarbonate type III H cages with a stainless-steel wire lid and a polysulphide filtertop cage (Tecniplast, West Chester, PA, USA), with corncob and carboard tubes as bedding materials. An artificial light/dark cycle with a sequence of 12 h was applied. The room was ventilated with about 15–20 air changes/h. A temperature of 22 ± 2 °C and a relative humidity of $55 \pm 15\%$ was maintained.

Feed and water were provided ad libitum to all animals during the experiment. They received a commercial 2014S Teklad rodent diet (Teklad diets, ENVIGO) based on 14.3% crude protein, 4% fat, 48% carbohydrates and 4.1% crude fiber in addition to vitamins and fatty acids.

2.7.2. Test Substance and Dosing Concentration

Given the high viscosity of BNC aqueous suspensions, the concentration for daily gavage was adjusted to 1% (*m/v*), as measured by gravimetry after drying overnight at 105 °C, by adding dH_2O. The ground BNC was sterilized by autoclaving for 20 min at 120 °C and 1 bar; after cooling to RT, the suspensions were stored at 4 °C under sterile conditions. Before feeding, the suspensions were warmed to RT and homogenized by vigorously vortex agitation.

2.7.3. In Vivo Study

This study comprised one dose group (each animal received 0.75 mL of the previously prepared 1% BNC suspension daily for a 21-day period) of 4 animals (two male and two female). Oral gavage was performed in the morning at a fixed time, using a polypropylene gavage needle of 1.3×1.3 mm without a ball tip. Each animal was observed daily for clinical signs. As control, animals fed only with commercial feed for 21 days were used.

Following sacrifice, the animals' intestinal tract was collected and processed by the Swiss roll technique (SRT). The small intestines were collected and cut into three equal segments to obtain the duodenum, jejunum and ileum regions. The lumen of each small/large intestine portion was washed with PBS, processed by the SRT, cut longitudinally, opened so that the lumen is facing upward and then rolled. All of these tissues were embedded in OCT compound (Tissue-Tek®, SakuraTM, Flemingweg, The Netherlands), frozen and cryo-sectioned in 20 µm-thick slices (LEICA CM 1900, Wetzlar, Germany).

2.7.4. Staining and Microscopic Observations

All samples were subjected to UV pre-treatment and final Sudan Black B (SBB) (Sigma-Aldrich, St. Louis, MO, USA) staining to remove tissue autofluorescence that could hinder cellulose detection with fluorescent CBM [48]. Briefly, after samples' pre-treatment with UV for 2 h, they were fixed with 4% PFA for 30 min, followed by a washing step with PBS buffer at RT. Cells' permeabilization was carried out with 0.5% Triton X-100 in PBS at RT, in a humid chamber (HC), then the blocking step was performed with 10% (*v/v*) fetal calf serum (FCS) (Sigma-Aldrich) in PBS for 1 h. For nucleus visualization, the slides were incubated with DAPI (1 µg/mL) for 10 min. After washing, pha-red (1 µg/mL) was applied, for 30 min in a HC, to stain the actin cytoskeleton. Then, samples were washed with PBS (three times) and incubated with GFP–CBM (0.05 mg/mL) for 2 h at RT in a HC. Sections were washed again with PBS and incubated with 0.1% SBB for 20 min in HC. Finally, the slides were washed with PBS and mounted with permafluor mounting media (Sigma-Aldrich, St. Louis, MO, USA).

At last, FM images of the stained histological samples were acquired using the Oympus BX61 Confocal Scanning Laser Microscope (Tokyo, Japan).

2.8. Statistical Analysis

The obtained raw data were statistically analyzed using GraphPad Prism software (Version 8.0.2.263, Graph Pad Software, Inc, San Diego, CA, USA). Differences in data from cytotoxicity assay were analyzed statistically using one-way ANOVA, with Dunnett's multiple comparison test. All of the treatment conditions were compared with the control, and a 95% level of confidence ($p < 0.05$) was used.

3. Results and Discussion

The extent of BNC degradation in the human microbiome as well as its fate in the human body are still poorly characterized. This represents a primary obstacle towards its use in food systems, given the current regulatory constraints on the use of nanomaterials. In this work, we first aimed at establishing the methodological tools that allow the detection of individual BCNC through fluorescent microscopy. For this, BNC was chemically hydrolyzed into its elemental structural unit (BCNC). CLEM was then used to demonstrate that the same isolated BNC nanocrystals identified by TEM could also be visualized by FM. Secondly, using FM, we aimed to evaluate the cellular uptake of BCNC by phagocytic cells. Finally, an exploratory in vivo study using Wistar rats was performed to evaluate the behavior of BNC along the GIT, namely its potential absorption.

3.1. BCNC Production and Characterization

The combination of sulfuric and hydrochloric acids has been shown to allow the production of stable BCNC with high dispersibility in aqueous suspensions [39]. In this work, the surface charge of the prepared BCNC was evaluated by means of Zeta-potential analysis. It is generally considered that colloidal systems with particles bearing a modulus of >30 mV reflect good stability due to electrostatic repulsion forces [49]. BCNC presented an average zeta value of -45.2 ± 0.7, indicative of such good stability. The highly negative surface charge density of the BCNC is due to the conjugated sulfate groups generated from the esterification of the hydroxyl groups at the surface of the BCNC, since HCl does not react with the hydroxyl groups [49–51].

Figure 1 displays the FTIR spectra of BNC and BCNC. Both samples exhibited similar vibration bands, namely ~900, 1030–1165, 1375–1475, 1635, 2900 and 3350–3400 cm^{-1}, well described in the literature: C-O-C stretching from β-(1,4) glycosidic linkages are attributed to the band near 900 cm^{-1}; peaks ranging from 1030 to 1165 cm^{-1} have reflected other C-O bonds; the ones comprised between 1375–1435 cm^{-1} are assigned to C-H bending; the wide bands around 3350–3400 and 1630–1700 cm^{-1} were related to the O-H groups of cellulose; finally, the band at 2900 cm^{-1} represented C-H stretching [52]. As expected, the small shoulder at 811 cm^{-1} (black arrow), only observed in the BCNC spectra, corresponds to the C-O-S group vibration, owing to the establishment of sulfate esters on nanocrystal surfaces during the acid hydrolysis [52].

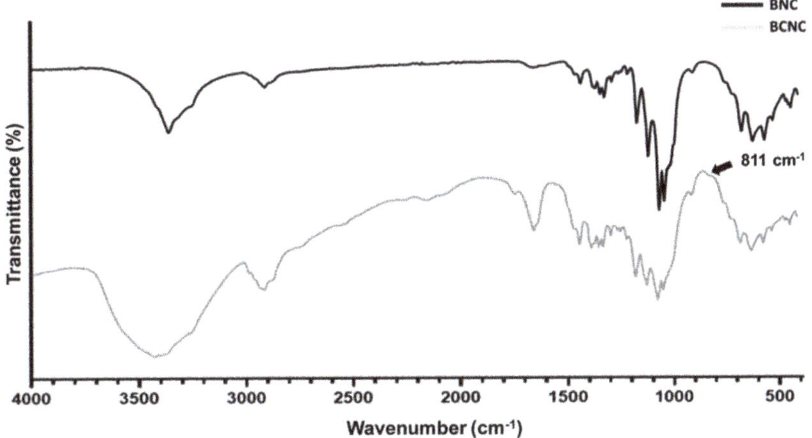

Figure 1. FTIR spectra of BNC and BCNC.

With TEM imaging (Figure 2), BCNC display a typical needle-shaped structure [46,48], with dimensions of 6–25 nm in width (average W of 13 ± 4 nm) and about 50–1100 nm in length (average L of 303 ± 199 nm). The BCNC dimensions are close to those reported in the literature for BCNC prepared under similar hydrolysis conditions [39,41,49].

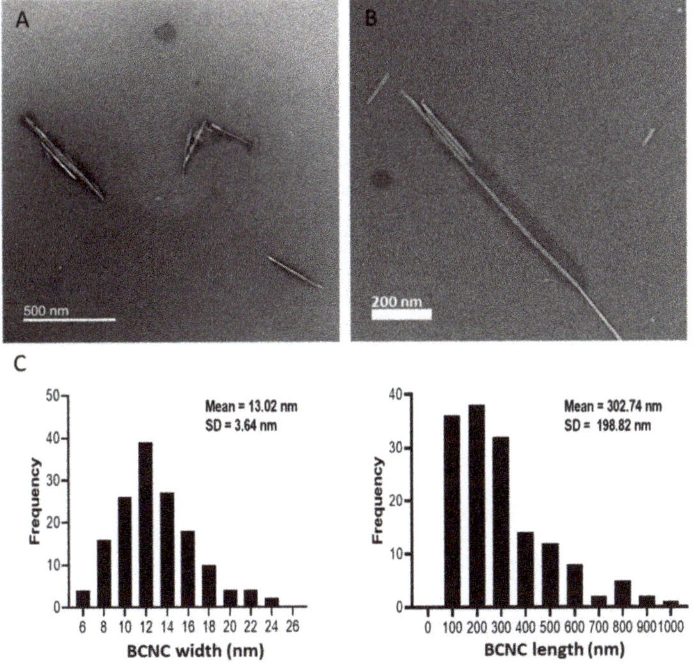

Figure 2. (**A**,**B**) TEM micrographs of the BCNC prepared by the acid hydrolysis of BNC and (**C**) the corresponding particle-size distributions. Scale bars: (**A**) 500 nm and (**B**) 200 nm.

As could be expected from their high modular value of the zeta potential, any aggregates observed consisted of only few nanocrystals, mostly organized 'side by side'. Similar size-distribution profiles were reported in other works, in particular for cellulose nanocrystals produced by HCl hydrolysis [53–55].

3.2. GFP–CBM Adsorption onto BNC and BCNC

A GFP-fused carbohydrate-binding module (type 3a family [41], GFP–CBM3A) was selected to detect cellulose in biological tissues and to track its localization in histological slides. The GFP–CBM adsorption on BNC and BCNC was compared and the obtained data fitted to the Langmuir adsorption model [43,44]. Figure 3 shows the adsorption isotherms of GFP–CBM onto BNC and BCNC and the respective parameters of the Langmuir isotherm.

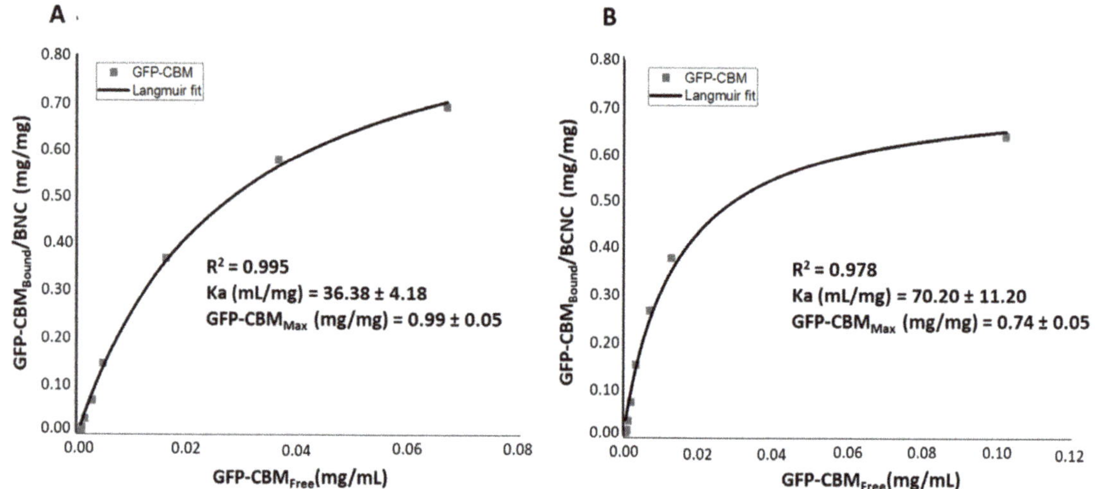

Figure 3. Adsorption isotherm of GFP–CBM onto: (**A**) BNC and (**B**) BCNC, after 2 h of incubation.

In both cases, the calculated R^2 values were higher than 0.97, indicating a good fitting of the experimental data to the Langmuir isotherm model, as also demonstrated in other studies using type-A CBMs towards micro- and nano-crystalline cellulose, either from plants or bacterial [38,56–59].

Regarding the Langmuir constant K_a, BCNC (~70 mg/mL) show a higher value than BNC (~36 mg/mL), suggesting the GFP–CBM affinity is higher in the former case. The higher surface area of BCNC relative to BNC is probably responsible for the higher (apparent) affinity in the former case. However, the maximum adsorption was lower for BCNC (0.74 mg/mL for BCNC vs. 0.99 mg/mL for BNC). The analysis of the results obtained is not straightforward, as the two celluloses likely present different surface areas, as well as different surface charges. Several structural studies showed that GFP–CBM from the type 3A family, such as the one used in this work, bind to the hydrophobic face of crystalline cellulose, notably on the 110 face of crystallite [58]. Thus, the adsorption in higher amounts to BNC could be expected, since the surface charge on BCNC does not favor the interaction. With regards to the main purpose of this work, we could conclude that the interaction of the GFP–CBM with BCNC is not hampered by some surface sulfation associated with hydrolysis. Additionally, as shown in Figure 4, FM allows the visualization of the GFP–CBM adsorbed on hydrolyzed nanocellulose.

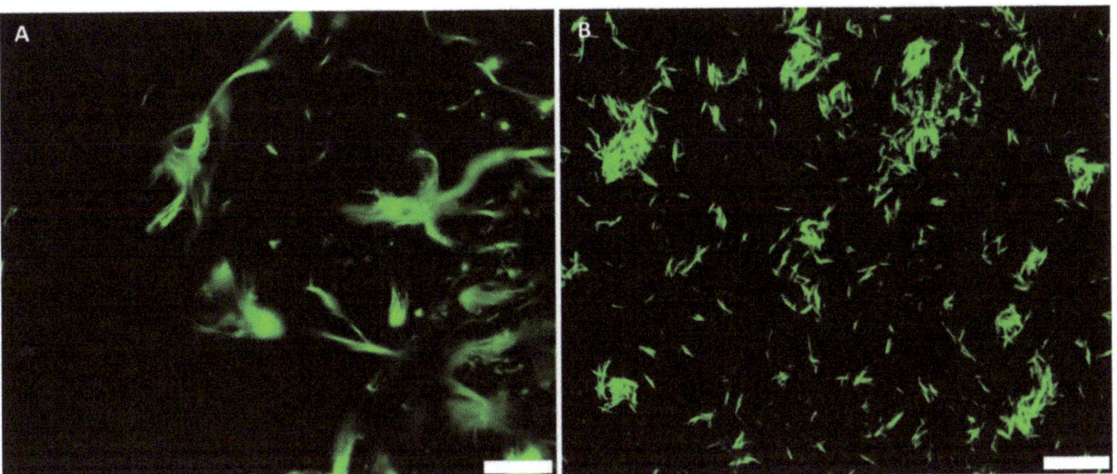

Figure 4. FM images of GFP–CBM (0.05 mg/mL) bound onto (**A**) BNC and (**B**) BCNC after 2 h of incubation. BNC and BCNC stock solutions were both at 0.25 mg/mL. Scale bar: 10 μm.

3.3. Considerations for Correlative Light Electron Micoscopy

In CLEM, FM observation is performed first, followed by TEM, because the electron beam can destroy the sample to some extent. The transition between FM and TEM must be made without dislocating the nanocrystals, a requirement strictly essential for any correlation to be possible. This proved to be challenging, since drying the sample before FM, to ensure the proper adhesion of the nanocrystals to the grid's surface, resulted in a dramatic reduction of the fluorescence. On the other hand, using a moisturized/wet sample for FM analysis, followed by drying prior to TEM, resulted in the dislocation of BCNC on the surface.

These issues have been overcome by using gelatin, which provides both steric and electrostatic stabilization of colloidal suspensions [60–62]. Thus, BCNC were dispersed in gelatin, and a certain suspension volume was placed on the surface of the grid. Ahmad et al. (2011) demonstrated that silver nanoparticles synthesized in edible-grade gelatin showed a more homogenous size distribution compared to those produced by conventional methods; also, the nanoparticles had lower levels of aggregation in gelatin than the dried ones [63]. In a similar way, we hypothesized that, (i) in gelatin, the BCNC should remain well dispersed and spatially stabilized upon drying and (ii) the gelatin matrix would not interfere in the visualization of BCNC either in FM (e.g., due to autofluorescence) or in TEM (e.g., by exhibiting nanofibers with a complex shape that could hinder the detection of BCNC).

As shown in Figure 5, the change of the dispersion matrix from water to gelatin resulted in well-dispersed GFP–CBM-labeled BCNC, as seen in FM (Figure 5A) and also in TEM (Figure 5B). While the dispersions of GFP–CBM:BCNC are stable, due to the sulfated groups, upon drying they tend to aggregate (Figure 5C), which is not the case when dispersed in gelatin (Figure 5B). Furthermore, in the latter, the calculated BCNC width (15 ± 4 nm) and length (449 ± 250 nm) were in accordance with the previous measurements obtained for BCNC in water suspensions (Figure 2C). Additionally, gelatin showed no autofluorescence, providing a suitable environment where nanocrystals exhibit high fluorescence. Thus, CLEM was performed using dispersions of GFP–CBM:BCNC in gelatin.

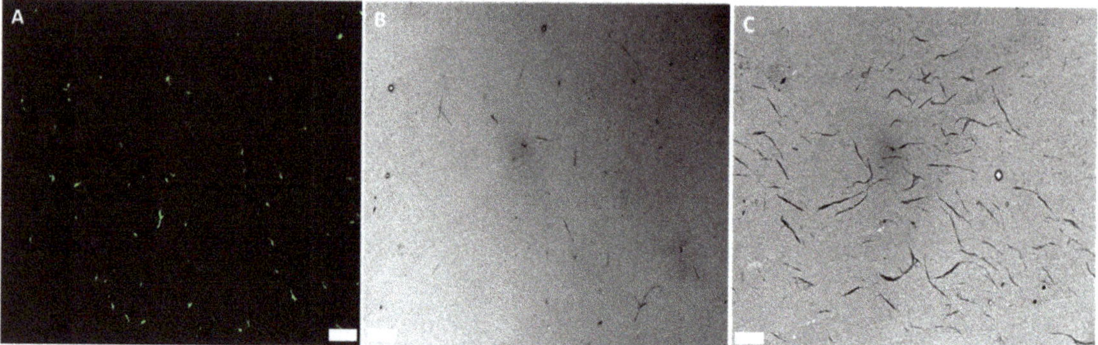

Figure 5. Microscopic images of GFP–CBM:BCNC (0.002 mg/mL) mixed with 2% gelatin (1:1) (**A**,**B**) and of GFP–CBM:BCNC aqueous suspension (**C**). (**A**) FM image. (**B**,**C**) TEM images. Scale bars: (**A**) 20 μm and (**B**,**C**) 1 μm.

3.4. Correlative Light Electron Microscopy

GFP–CBM: BCNC complexes were stabilized in gelatin, and a drop of the mixture was placed on a grid in order to assess the sensitivity of FM to detecting nanocrystals by CLEM. Given its lower resolution (limit ~200 nm [64]) compared to TEM, FM alone does not allow the measurement of the size of BCN crystals that give rise to the green signal detected (Figure 6), and it is not possible to reach conclusions on whether the fluorescent signal is generated by single crystals or by aggregates of several crystals. However, with CLEM imaging, it is possible to confirm the co-localization of the signals observed by FM and TEM, and, in several cases, the signal detected by FM seems indeed to correspond to single crystals, as detected by TEM. These results show that, despite the lower resolution limit compared to TEM, fluorescence imaging can be used as a feasible, robust and highly specific technique for the detection of isolated cellulose nanocrystals and thus also of larger fragments of BNC or smaller fibers released in vivo, in biological tissues.

3.5. In Vitro Assays

3.5.1. Cytotoxicity Evaluation

Evaluating the biocompatibility of a material is an essential step toward its acceptance. Cell culture studies usually are the first step for the biocompatibility evaluation, as they are simple systems that minimize the effect of other variables. Although BNC is regarded as a biocompatible material [65], there are not so many reports on the cytotoxicity of BCNC, which we intend to use while studying the internalization (and subsequent detection) by macrophages. In this way, the cytotoxicity of BCNC was characterized, using both fibroblasts and macrophages.

Results from the metabolic viability assay of L929 and BMMΦ cells when cultivated with BCNC for 24 h (Figure 7) show that BCNC did not significantly affect the viability of either cell type, even at very high concentrations (1 and 2 mg/mL, which are unlikely to be reached in vivo). The obtained data show that BCNC are not cytotoxic—as defined by ISO 10993, they did not reduce cell viability by more than 30%—consistent with reports from the literature [38]. However, a significant increase (by 50% on average) in the metabolic viability was obtained in BMMΦ cells after treatment with all tested concentrations. The effect was previously observed by others when exposing macrophages to vegetable nano and microfibrillated cellulose [66,67].

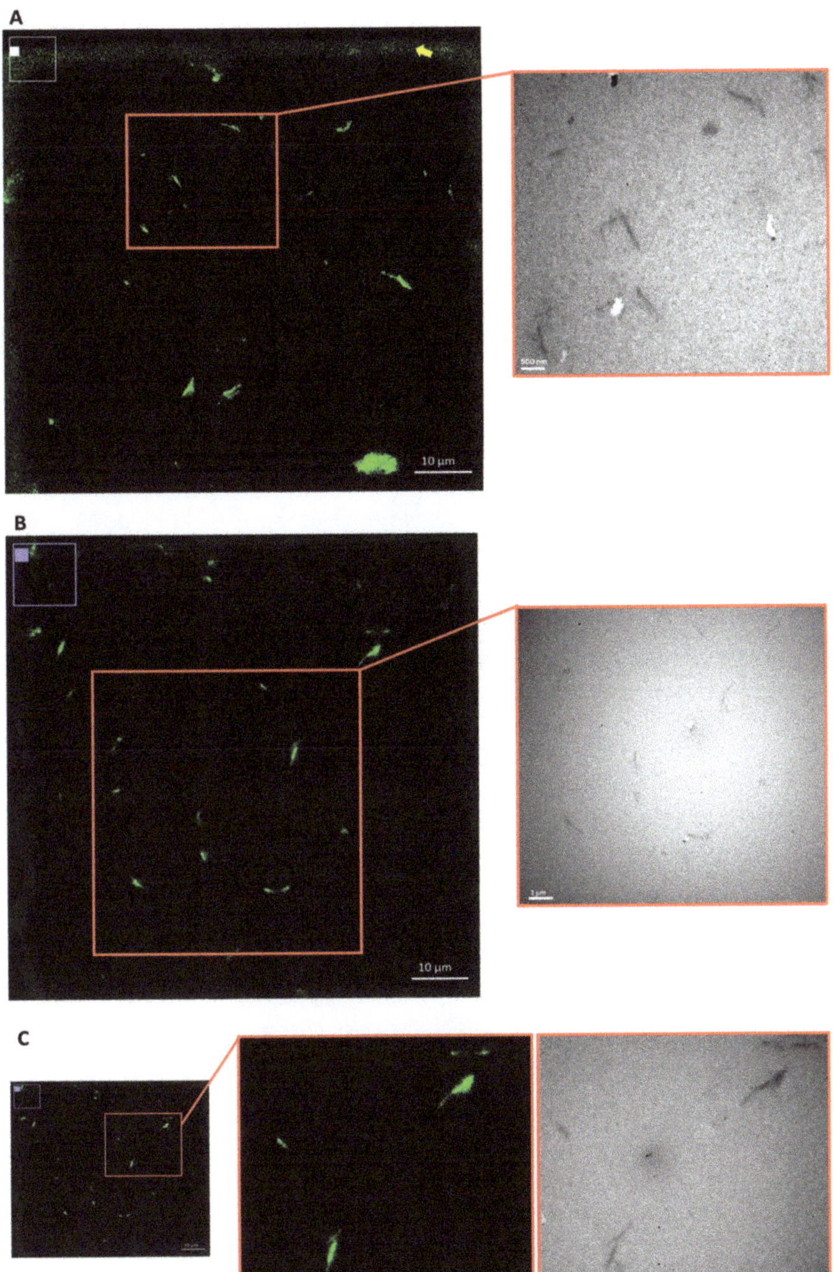

Figure 6. CLEM images of BCNC labeled with GFP–CBM. Small green dots in FM correspond to single BCNC in TEM (highlighted in the red boxes). The solid part of the TEM grid had some autofluorescence (yellow arrow); however, it did not interfere with the visualization as the GFP–CBM:BCNC-gelatin mixture was fixed in the transparent network mesh. Scale bars: (**A–C**) 10 µm for FM, (**A**,**C**) 500 nm and (**B**) 1 µm for TEM.

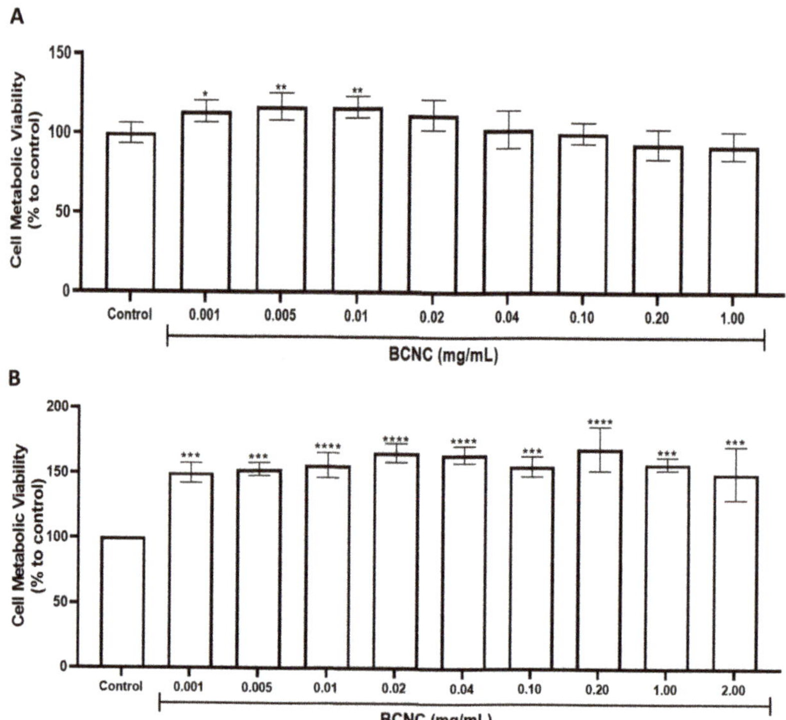

Figure 7. Metabolic viability of (**A**) L929 and (**B**) BMMΦ cells when cultured with increasing doses of BCNC for 24 h, as assessed by the Resazurin assay. PBS was used as the control. Data is expressed as a percentage relative to the control and is the mean ± SD of three independent experiments. All treatment conditions were compared with the control using Dunnett's multiple comparison test. * $p < 0.05$, ** $p < 0.01$, *** $p < 0.001$ and **** $p < 0.0001$ compared to control (non-treated cells).

3.5.2. Uptake of BCNC by BMMΦ Primary Cells

As in most tissues, gut-resident macrophages are important immune sentinels and effector populations. Positioned in close apposition to the epithelial layer, they are able to rapidly uptake and respond to any material breaching this barrier [68]. Thus, the process of the internalization of BCNC by macrophages was evaluated over time using GFP–CBM as described above.

Fluorescence images (Figure 8 showed that during short incubation periods (2 to 6 h), macrophages were already able to internalize BCNC; however, the amount of internalized material was very small. Despite this, it was possible to see labeled cellulose crystals inside some cells (most of the non-internalized BCNC was washed before cell fixation). At 24 h, clearly, a higher amount of the nanomaterial was uptaken. Several studies have reported that the macrophage internalization of foreign particles increases over time [69,70]. For instance, Erdem et al. (2021) showed that pristine cellulose nanocrystal uptake by alveolar macrophages increased significantly as the cell exposure period was longer (2, 6 and 24 h) [71]. Furthermore, we made an attempt to identify BCNC through ultrastructural analysis, which appear to be located within a lysosome (Supplementary Materials Figure S1). As described in the literature, macrophages are able to internalize particles with a diameter ranging from 6 to 6000 nm [72]. Therefore, we demonstrate that, using fluorescent CBM, we are able to detect BCNC inside the macrophages, and thus, this may also be possible in vivo.

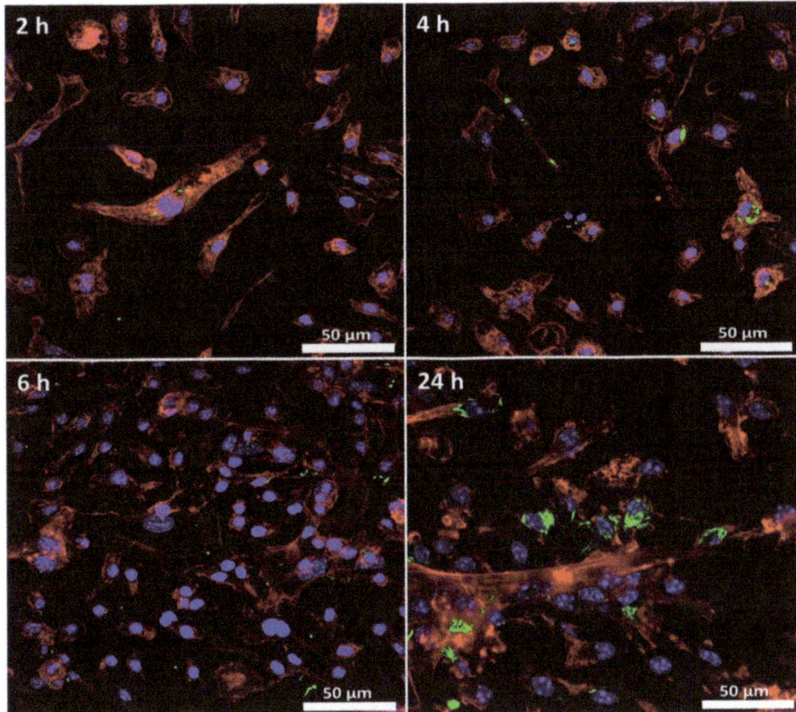

Figure 8. Cellular uptake of BCNC by phagocytic cells. FM images of macrophages exposed to 0.001 mg/mL BCNC for 2, 4, 6 and 24 h. DAPI (blue), Pha-red (red) and GFP–CBM (green) were used for nuclei, actin cytoskeleton and BCNC visualization, respectively. Scale bar: 50 µm.

3.6. *Nanocellulose Tracking in Animal Tissues*

Despite several studies having reported important data supporting the safety of BNC as a food additive [33], its fate in the human body upon ingestion is still not totally clarified. Indeed, detecting small materials in biological tissues through imaging techniques is often hampered either by high background levels or by the lack of sensitive, nanomaterial-specific detection methods. Once it was demonstrated that fluorescence-based detection is a feasible method for the detection of nano-sized cellulose particles, a 21-day in vivo pilot study was made, to determine whether the same methodology was suitable for detecting micro- and/or nanocellulose in histological samples of animal tissues. It should be noted that UV + SBB treatment [48] eliminated the autofluorescence associated with this type of tissue, which, in several studies, made it difficult to apply fluorescence-based detection methods [73].

To better assess whether GFP–CBM-based detection is able to detect trace amount of cellulose along the GIT, the rats' intestines were washed to remove, as much as possible, the cellulose present in the lumen and processed with SRT. Cellulose was found only in a few cases in the intestinal lumen or between the villi (Figure 9), without any notable difference in frequency or location among animals of different genders. However, no signs of the intestinal persorption of cellulose were found.

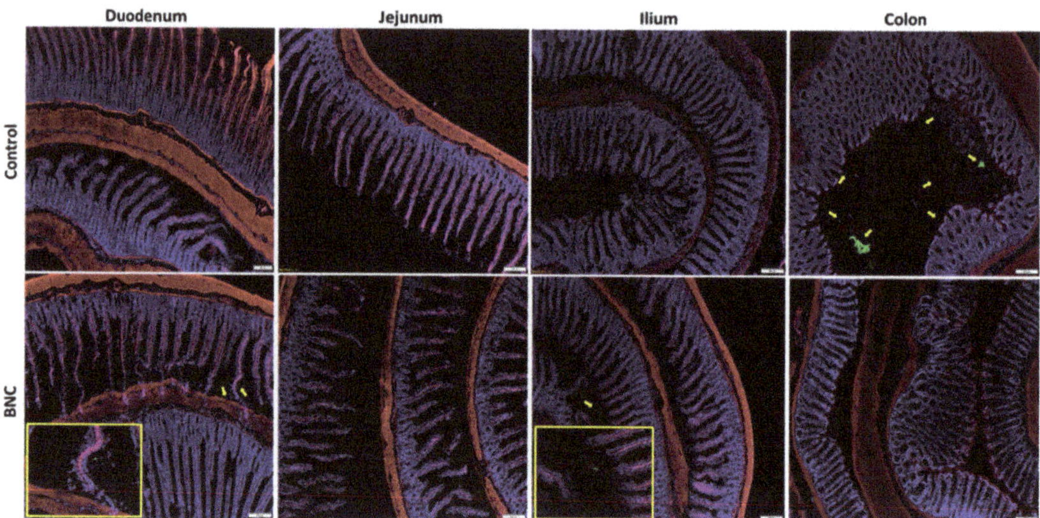

Figure 9. FM images of a male rat histological sample after a 21-day repeated-dose study. Four GIT regions (duodenum, jejunum, ilium and colon) were stained for nuclei (blue), actin cytoskeleton (red) and cellulose (green). Control (no gavage) and BNC (oral gavage) animals. Cellulose fibers were detected at the intestinal lumen or trapped between the villi (pointed out by the yellow arrow and the respective amplified image in the yellow box). Scale bar: 200 μm.

In fact, other studies examining the fate of cellulose particles in the intestinal tract have found no signs of translocation [26,33,74]. For instance, Mackie et al. (2019) sequentially studied the GIT fate of cellulose nanocrystal emulsions and their exposure to the intestinal mucus layer. They determined that the nanocrystals were clearly trapped in the intestinal mucosa, unable to reach the underlying epithelium, and were therefore considered safe emulsifiers [74]. Accordingly, the proposed routes of cellulose particle uptake predicted by others along the GIT, such as through M cells, the paracellular pathway and through enterocytes by transcytosis or passive diffusion [75], appear not to be supported since particles must first pass through the intestine mucus layer to subsequently interact with the gut wall. Our findings seem to confirm these reports, but a larger number of animals and a more detailed screening of the tissues must be performed to take more definitive conclusions.

4. Conclusions

A methodology for the specific detection of nano-scalar cellulose is critical for an understanding of its distribution in the human body. Herein, we demonstrate that fluorescence-based method can be used to detect and visualize different kinds of cellulose fibers, including nano-sized ones. The screening assay revealed simple, feasible and specific, allowing the detection of nanocrystals internalized in macrophages.

An exploratory work in vivo was performed, whereby no evidence of mucus layer translocation in the intestine was found. A more comprehensive study is required in order to take conclusions with regards to the potential cellulose persorption and thus contribute to the analysis of the safety of BNC as a food additive.

Supplementary Materials: The following supporting information can be downloaded at: https://www.mdpi.com/article/10.3390/nano12152605/s1, Figure S1: Uptake of BCNC by BMMΦ primary cells—Electron Microscopy.

Author Contributions: Conceptualization, M.G.; Formal analysis, R.M.; Investigation, R.M., A.C.R., R.S.-C., L.C., D.M. and P.S.; Methodology, R.M., A.C.R., R.S.-C., L.C. and D.M.; Project administration, P.S.; Supervision, F.D. and M.G.; Validation, F.D. and M.G.; Writing—original draft, R.M.; Writing—review & editing, P.S., F.D. and M.G. All authors have read and agreed to the published version of the manuscript.

Funding: This work was supported by the Portuguese Foundation for Science and Technology (FCT) under the scope of the strategic funding of UID/BIO/04469/2013 unit and COMPETE 2020 (POCI-01-0145-FEDER-006684) and BioTecNorte operation (NORTE-01-0145-FEDER-000004) funded by the European Regional Development Fund under the scope of Norte2020—Programa Operacional Regional do Norte.

Institutional Review Board Statement: Not applicable.

Informed Consent Statement: Not applicable.

Data Availability Statement: The relevant data for the discussion are completely supplied in the Results and Discussion section. Raw data are available upon request.

Acknowledgments: The authors acknowledge the support of i3S Scientific Platforms: (i) Advanced Light Microscopy and (ii) HEMS, both members of the national infrastructure PPBI-Portuguese Platform of BioImaging (supported by POCI-01-0145-FEDER-022122).

Conflicts of Interest: The authors declare no conflict of interest.

References

1. Ma, T.; Hu, X.; Lu, S.; Liao, X.; Song, Y.; Hu, X. Nanocellulose: A promising green treasure from food wastes to available food materials. *Crit. Rev. Food Sci. Nutr.* **2022**, *62*, 989–1002. [CrossRef]
2. Serpa, A.; Velásquez-Cock, J.; Gañán, P.; Castro, C.; Vélez, L.; Zuluaga, R. Vegetable nanocellulose in food science: A review. *Food Hydrocoll.* **2016**, *57*, 178–186. [CrossRef]
3. Endes, C.; Camarero-Espinosa, S.; Mueller, S.; Foster, E.J.; Petri-Fink, A.; Rothen-Rutishauser, B.; Weder, C.; Clift, M.J.D. A critical review of the current knowledge regarding the biological impact of nanocellulose. *J. Nanobiotechnol.* **2016**, *14*, 1–14. [CrossRef] [PubMed]
4. Da Gama, F.M.P.; Dourado, F. Bacterial NanoCellulose: What future? *Bioimpacts* **2018**, *8*, 1. [CrossRef] [PubMed]
5. Thomas, P.; Duolikun, T.; Rumjit, N.P.; Moosavi, S.; Lai, C.W.; Johan, M.R.B.; Fen, L.B. Comprehensive review on nanocellulose: Recent developments, challenges and future prospects. *J. Mech. Behav. Biomed. Mater.* **2020**, *110*, 103884. [CrossRef] [PubMed]
6. Dufresne, A. Nanocellulose processing properties and potential applications. *Curr. For. Rep.* **2019**, *5*, 76–89. [CrossRef]
7. Statista. Available online: https://www.statista.com/statistics/1192542/global-nanocellulose-market-size/ (accessed on 12 April 2022).
8. EFSA Panel on Food Additives and Nutrient Sources Added to Food (ANS); Younes, M.; Aggett, P.; Aguilar, F.; Crebelli, R.; Di Domenico, A.; Dusemund, B.; Filipič, M.; Frutos, M.J.; Galtier, P.; et al. Re-evaluation of celluloses E 460 (i), E 460 (ii), E 461, E 462, E 463, E 464, E 465, E 466, E 468 and E 469 as food additives. *EFSA J.* **2018**, *16*, e05047. [CrossRef]
9. Dourado, F.; Leal, M.; Martins, D.; Fontão, A.; Rodrigues, A.C.; Gama, M. Celluloses as food ingredients/additives: Is there a room for BNC? In *Bacterial Nanocellulose*; Gama, M., Dourado, F., Bielecki, S., Eds.; Elsevier: Amsterdam, The Netherlands, 2016; pp. 123–133. [CrossRef]
10. Turbak, A.F.; Snyder, F.W.; Sandberg, K.R. Microfibrillated cellulose, a new cellulose product: Properties, uses, and commercial potential. *J. Appl. Polym. Sci. Appl. Polym. Symp.* **1983**, *37*, 815–827.
11. DeLoid, G.M.; Sohal, I.S.; Lorente, L.R.; Molina, R.M.; Pyrgiotakis, G.; Stevanovic, A.; Zhang, R.; McClements, D.J.; Geitner, N.K.; Bousfield, D.W.; et al. Reducing intestinal digestion and absorption of fat using a nature-derived biopolymer: Interference of triglyceride hydrolysis by nanocellulose. *ACS Nano* **2018**, *12*, 6469–6479. [CrossRef]
12. Tarhan, Ö. Safety and regulatory issues of nanomaterials in foods. In *Handbook of Food Nanotechnology*; Jafari, S.M., Ed.; Academic Press: Cambrige, MA, USA, 2020; pp. 655–703. [CrossRef]
13. EFSA Scientific Committee; Hardy, A.; Benford, D.; Halldorsson, T.; Jeger, M.J.; Knutsen, H.K.; More, S.; Naegeli, H.; Noteborn, H.; Ockleford, C.; et al. Guidance on risk assessment of the application of nanoscience and nanotechnologies in the food and feed chain: Part 1, human and animal health. *EFSA J.* **2018**, *16*, e05327. [CrossRef]
14. Stoudmann, N.; Schmutz, M.; Hirsch, C.; Nowack, B.; Som, C. Human hazard potential of nanocellulose: Quantitative insights from the literature. *Nanotoxicology* **2020**, *14*, 1241–1257. [CrossRef]
15. Lu, Q.; Yu, X.; Yagoub, A.E.A.; Wahia, H.; Zhou, C. Application and challenge of nanocellulose in the food industry. *Food Biosci.* **2021**, *43*, 101285. [CrossRef]
16. Singh, G.; Saquib, S.; Gupta, A. Environmental, legal, health, and safety issue of nanocellulose. In *Nanocellulose Materials*; Oraon, R., Raetani, D., Singh, P., Hussain, C.M., Eds.; Elsevier: Amsterdam, The Netherlands, 2022; pp. 265–288.

17. Hillyer, J.F.; Albrecht, R.M. Gastrointestinal persorption and tissue distribution of differently sized colloidal gold nanoparticles. *J. Pharm. Sci.* **2021**, *90*, 1927–1936. [CrossRef] [PubMed]
18. Schleh, C.; Semmler-Behnke, M.; Lipka, J.; Wenk, A.; Hirn, S.; Schäffler, M.; Schmid, G.; Simon, U.; Kreyling, W.G. Size and surface charge of gold nanoparticles determine absorption across intestinal barriers and accumulation in secondary target organs after oral administration. *Nanotoxicology* **2012**, *6*, 36–46. [CrossRef]
19. Kim, K.S.; Suzuki, K.; Cho, H.; Youn, Y.S.; Bae, Y.H. Oral nanoparticles exhibit specific high-efficiency intestinal uptake and lymphatic transport. *ACS Nano* **2018**, *12*, 8893–8900. [CrossRef] [PubMed]
20. Tortella, G.R.; Rubilar, O.; Durán, N.; Diez, M.C.; Martínez, M.; Parada, J.; Seabra, A.B. Silver nanoparticles: Toxicity in model organisms as an overview of its hazard for human health and the environment. *J. Hazard. Mater.* **2020**, *390*, 121974. [CrossRef]
21. Chong, C.L.; Fang, C.M.; Pung, S.Y.; Ong, C.E.; Pung, Y.F.; Kong, C.; Pan, Y. Current updates on the in vivo assessment of zinc oxide nanoparticles toxicity using animal models. *Bionanoscience* **2021**, *11*, 590–620. [CrossRef]
22. Brohi, R.D.; Wang, L.; Talpur, H.S.; Wu, D.; Khan, F.A.; Bhattarai, D.; Rehman, Z.-U.; FarmanUllah, F.; Huo, L.-J. Toxicity of nanoparticles on the reproductive system in animal models: A review. *Front. Pharmacol.* **2017**, *8*, 606. [CrossRef]
23. Juhr, N.C.; Franke, J. A method for estimating the available energy of incompletely digested carbohydrates in rats. *J. Nutr.* **1992**, *122*, 1425–1433. [CrossRef] [PubMed]
24. Huang, Y.; Zhu, C.; Yang, J.; Nie, Y.; Chen, C.; Sun, D. Recent advances in bacterial cellulose. *Cellulose* **2014**, *21*, 1–30. [CrossRef]
25. Joint FAO/WHO Expert Committee on Food Additives; WHO Expert Committee on Food Additives. *Safety Evaluation of Certain Food Additives and Contaminants*; WHO Press: Geneva, Switzerland, 2002.
26. Kotkoskie, L.A.; Butt, M.T.; Selinger, E.; Freeman, C.; Weiner, M.L. Qualitative investigation of uptake of fine particle size microcrystalline cellulose following oral administration in rats. *J. Anat.* **1996**, *189*, 531. [PubMed]
27. Nagele, W.; Müller, N.; Brugger-Pichler, E.; Nagele, J. Persorption Of Plant Microparticles After Oral Plant Food Intake. *Int. J. Herb. Med.* **2013**, *2*, 1–8.
28. Kelleher, J.; Walters, M.P.; Srinivasan, T.R.; Hart, G.; Findlay, J.M.; Losowsky, M.S. Degradation of cellulose within the gastrointestinal tract in man. *Gut* **1984**, *25*, 811–815. [CrossRef]
29. Winkler, H.C.; Suter, M.; Naegeli, H. Critical review of the safety assessment of nano-structured silica additives in food. *J. Nanobiotechnol.* **2016**, *14*, 1–9. [CrossRef] [PubMed]
30. Jafarizadeh-Malmiri, H.; Sayyar, Z.; Anarjan, N.; Berenjian, A. Nano-additives for food industries. In *Nanobiotechnology in Food: Concepts, Applications and Perspectives*; Springer: Cham, Switzerland, 2019; pp. 41–68. [CrossRef]
31. Blasco, C.; Pico, Y. Determining nanomaterials in food. *TrAC Trends Anal. Chem.* **2011**, *30*, 84–99. [CrossRef]
32. Dasgupta, N.; Ranjan, S. Nano-Food Toxicity and Regulations. In *An Introduction to Food Grade Nanoemulsions*; Environmental Chemistry for a Sustainable World; Springer: Singapore, 2018; pp. 151–179. [CrossRef]
33. Dourado, F.; Gama, M.; Rodrigues, A.C. A review on the toxicology and dietetic role of bacterial cellulose. *Toxicol. Rep.* **2017**, *4*, 543–553. [CrossRef]
34. Cortese, K.; Diaspro, A.; Tacchetti, C. Advanced correlative light/electron microscopy: Current methods and new developments using Tokuyasu cryosections. *J. Histochem. Cytochem.* **2009**, *57*, 1103–1112. [CrossRef] [PubMed]
35. Timmermans, F.J.; Otto, C. Contributed review: Review of integrated correlative light and electron microscopy. *Rev. Sci. Instrum.* **2015**, *86*, 011501. [CrossRef] [PubMed]
36. Sjollema, K.A.; Schnell, U.; Kuipers, J.; Kalicharan, R.; Giepmans, B.N. Correlated light microscopy and electron microscopy. *Methods Cell Biol.* **2012**, *111*, 157–173. [CrossRef] [PubMed]
37. De Boer, P.; Hoogenboom, J.P.; Giepmans, B.N. Correlated light and electron microscopy: Ultrastructure lights up! *Nat. Methods* **2015**, *12*, 503–513. [CrossRef] [PubMed]
38. Ventura, C.; Pinto, F.; Lourenço, A.F.; Ferreira, P.J.; Louro, H.; Silva, M.J. On the toxicity of cellulose nanocrystals and nanofibrils in animal and cellular models. *Cellulose* **2020**, *27*, 5509–5544. [CrossRef]
39. Vasconcelos, N.F.; Feitosa, J.P.A.; da Gama, F.M.P.; Morais, J.P.S.; Andrade, F.K.; de Souza, M.D.S.M.; de Freitas Rosa, M. Bacterial cellulose nanocrystals produced under different hydrolysis conditions: Properties and morphological features. *Carbohydr. Polym.* **2017**, *155*, 425–431. [CrossRef] [PubMed]
40. Blanco, A.; Monte, M.C.; Campano, C.; Balea, A.; Merayo, N.; Negro, C. Nanocellulose for industrial use: Cellulose nanofibers (CNF), cellulose nanocrystals (CNC), and bacterial cellulose (BC). In *Handbook of Nanomaterials for Industrial Applications*; Hussain, C.M., Ed.; Elsevier: Amsterdam, The Netherlands, 2018; pp. 74–126. [CrossRef]
41. Knudsen, K.B.; Kofoed, C.; Espersen, R.; Højgaard, C.; Winther, J.R.; Willemoës, M.; Wedin, I.; Nuopponen, M.; Vilske, S.; Aimonen, K.; et al. Visualization of nanofibrillar cellulose in biological tissues using a biotinylated carbohydrate binding module of β-1, 4-Glycanase. *Chem. Res. Toxicol.* **2015**, *28*, 1627–1635. [CrossRef]
42. Rasband, W.S.; ImageJ, U.S. National Institutes of Health, Bethesda, MD, USA. 1997–2022. Available online: https://imagej.nih.gov/ij/ (accessed on 25 July 2022).
43. Kim, D.W.; Jang, Y.H.; Kim, C.S.; Lee, N.S. Effect of metal ions on the degradation and adsorption of two cellobiohydrolases on microcrystalline cellulose. *Bull. Korean Chem. Soc.* **2001**, *22*, 716–720. [CrossRef]
44. Liu, C.H.; Tsao, M.H.; Sahoo, S.L.; Wu, W.C. Magnetic nanoparticles with fluorescence and affinity for DNA sensing and nucleus staining. *RSC Adv.* **2017**, *7*, 5937–5947. [CrossRef]

45. Origin(Pro), Version 9.5.1.195. OriginLab Corporation: Northampton, MA, USA. Available online: https://www.originlab.com/ (accessed on 25 July 2022).
46. Silva-Carvalho, R.; Fidalgo, J.; Melo, K.R.; Queiroz, M.F.; Leal, S.; Rocha, H.A.; Cruz, T.; Parpot, P.; Tomás, A.; Gama, M. Development of dextrin-amphotericin B formulations for the treatment of Leishmaniasis. *Int. J. Biol. Macromol.* **2020**, *153*, 276–288. [CrossRef] [PubMed]
47. O'brien, J.; Wilson, I.; Orton, T.; Pognan, F. Investigation of the Alamar Blue (resazurin) fluorescent dye for the assessment of mammalian cell cytotoxicity. *Eur. J. Biochem.* **2000**, *267*, 5421–5426. [CrossRef]
48. Viegas, M.S.; Martins, T.C.; Seco, F.; Do Carmo, A. An improved and cost-effective methodology for the reduction of autofluorescence in direct immunofluorescence studies on formalin-fixed paraffin-embedded tissues. *Eur. J. Histochem.* **2007**, *51*, 59–66.
49. Mirhosseini, H.; Tan, C.P.; Hamid, N.S.; Yusof, S. Effect of Arabic gum, xanthan gum and orange oil contents on ζ-potential, conductivity, stability, size index and pH of orange beverage emulsion. *Colloids Surf. A Physicochem. Eng. Asp.* **2008**, *315*, 47–56. [CrossRef]
50. Campano, C.; Balea, A.; Blanco, Á.; Negro, C. A reproducible method to characterize the bulk morphology of cellulose nanocrystals and nanofibers by transmission electron microscopy. *Cellulose* **2020**, *27*, 4871–4887. [CrossRef]
51. Roman, M.; Winter, W.T. Effect of sulfate groups from sulfuric acid hydrolysis on the thermal degradation behavior of bacterial cellulose. *Biomacromolecules* **2004**, *5*, 1671–1677. [CrossRef]
52. Listyanda, R.F.; Wildan, M.W.; Ilman, M.N. Preparation and characterization of cellulose nanocrystal extracted from ramie fibers by sulfuric acid hydrolysis. *Heliyon* **2020**, *6*, e05486. [CrossRef]
53. Shang, Z.; An, X.; Seta, F.T.; Ma, M.; Shen, M.; Dai, L.; Liu, H.; Ni, Y. Improving dispersion stability of hydrochloric acid hydrolyzed cellulose nano-crystals. *Carbohydr. Polym.* **2019**, *222*, 115037. [CrossRef] [PubMed]
54. Vanderfleet, O.M.; Cranston, E.D. Production routes to tailor the performance of cellulose nanocrystals. *Nat. Rev. Mater.* **2021**, *6*, 124–144. [CrossRef]
55. Liu, L.; Kong, F. The behavior of nanocellulose in gastrointestinal tract and its influence on food digestion. *J. Food Eng.* **2021**, *292*, 110346. [CrossRef]
56. Ong, E.; Gilkes, N.R.; Miller, R.C., Jr.; Warren, R.A.J.; Kilburn, D.G. The cellulose-binding domain (CBDCex) of an exoglucanase from Cellulomonas fimi: Production in Escherichia coli and characterization of the polypeptide. *Biotechnol. Bioeng.* **1993**, *42*, 401–409. [CrossRef]
57. Boraston, A.B.; Bolam, D.N.; Gilbert, H.J.; Davies, G.J. Carbohydrate-binding modules: Fine-tuning polysaccharide recognition. *Biochem. J.* **2004**, *382*, 769–781. [CrossRef] [PubMed]
58. Lehtiö, J.; Sugiyama, J.; Gustavsson, M.; Fransson, L.; Linder, M.; Teeri, T.T. The binding specificity and affinity determinants of family 1 and family 3 cellulose binding modules. *Proc. Natl. Acad. Sci. USA* **2003**, *100*, 484–489. [CrossRef]
59. Creagh, A.L.; Ong, E.; Jervis, E.; Kilburn, D.G.; Haynes, C.A. Binding of the cellulose-binding domain of exoglucanase Cex from Cellulomonas fimi to insoluble microcrystalline cellulose is entropically driven. *Proc. Natl. Acad. Sci. USA* **1996**, *93*, 12229–12234. [CrossRef] [PubMed]
60. Nakagawa, K.; Surassmo, S.; Min, S.G.; Choi, M.J. Dispersibility of freeze-dried poly (epsilon-caprolactone) nanocapsules stabilized by gelatin and the effect of freezing. *J. Food Eng.* **2011**, *102*, 177–188. [CrossRef]
61. Anarjan, N.; Nehdi, I.A.; Sbihi, H.M.; Al-Resayes, S.I.; Malmiri, H.J.; Tan, C.P. Preparation of astaxanthin nanodispersions using gelatin-based stabilizer systems. *Molecules* **2014**, *19*, 14257–14265. [CrossRef] [PubMed]
62. Likos, C.N.; Vaynberg, K.A.; Löwen, H.; Wagner, N.J. Colloidal stabilization by adsorbed gelatin. *Langmuir* **2000**, *16*, 4100–4108. [CrossRef]
63. Bin Ahmad, M.; Lim, J.J.; Shameli, K.; Ibrahim, N.A.; Tay, M.Y. Synthesis of silver nanoparticles in chitosan, gelatin and chitosan/gelatin bionanocomposites by a chemical reducing agent and their characterization. *Molecules* **2011**, *16*, 7237–7248. [CrossRef]
64. Cox, G.; Sheppard, C.J. Practical limits of resolution in confocal and non-linear microscopy. *Microsc. Res. Tech.* **2004**, *63*, 18–22. [CrossRef]
65. Kim, G.D.; Yang, H.; Park, H.R.; Park, C.S.; Park, Y.S.; Lee, S.E. Evaluation of immunoreactivity of in vitro and in vivo models against bacterial synthesized cellulose to be used as a prosthetic biomaterial. *Biochip J.* **2013**, *7*, 201–209. [CrossRef]
66. Lopes, V.R.; Sanchez-Martinez, C.; Strømme, M.; Ferraz, N. In vitro biological responses to nanofibrillated cellulose by human dermal, lung and immune cells: Surface chemistry aspect. *Part. Fibre Toxicol.* **2017**, *14*, 1–13. [CrossRef]
67. Vartiainen, J.; Pöhler, T.; Sirola, K.; Pylkkänen, L.; Alenius, H.; Hokkinen, J.; Tapper, U.; Lahtinen, P.; Kapanen, A.; Putkisto, K.; et al. Health and environmental safety aspects of friction grinding and spray drying of microfibrillated cellulose. *Cellulose* **2011**, *18*, 775–786. [CrossRef]
68. Grainger, J.R.; Konkel, J.E.; Zangerle-Murray, T.; Shaw, T.N. Macrophages in gastrointestinal homeostasis and inflammation. *Pflug. Arch. Eur. J. Physiol.* **2017**, *469*, 527–539. [CrossRef] [PubMed]
69. Mendes, R.G.; Mandarino, A.; Koch, B.; Meyer, A.K.; Bachmatiuk, A.; Hirsch, C.; Gemming, T.; Schmidt, O.G.; Liu, Z.; Rümmeli, M.H. Size and time dependent internalization of label-free nano-graphene oxide in human macrophages. *Nano Res.* **2017**, *10*, 1980–1995. [CrossRef]

70. Erdem, J.S.; Alswady-Hoff, M.; Ervik, T.K.; Skare, Ø.; Ellingsen, D.G.; Zienolddiny, S. Cellulose nanocrystals modulate alveolar macrophage phenotype and phagocytic function. *Biomaterials* **2019**, *203*, 31–42. [CrossRef]
71. Petithory, T.; Pieuchot, L.; Josien, L.; Ponche, A.; Anselme, K.; Vonna, L. Size-Dependent Internalization Efficiency of Macrophages from Adsorbed Nanoparticle-Based Monolayers. *Nanomaterials* **2021**, *11*, 1963. [CrossRef] [PubMed]
72. Gustafson, H.H.; Holt-Casper, D.; Grainger, D.W.; Ghandehari, H. Nanoparticle uptake: The phagocyte problem. *Nano Today* **2015**, *10*, 487–510. [CrossRef] [PubMed]
73. Wizenty, J.; Ashraf, M.I.; Rohwer, N.; Stockmann, M.; Weiss, S.; Biebl, M.; Pratschke, J.; Aigner, F.; Wuensch, T. Autofluorescence: A potential pitfall in immunofluorescence-based inflammation grading. *J. Immunol. Methods* **2018**, *456*, 28–37. [CrossRef] [PubMed]
74. Mackie, A.; Gourcy, S.; Rigby, N.; Moffat, J.; Capron, I.; Bajka, B. The fate of cellulose nanocrystal stabilised emulsions after simulated gastrointestinal digestion and exposure to intestinal mucosa. *Nanoscale* **2019**, *11*, 2991–2998. [CrossRef] [PubMed]
75. Torcello-Gómez, A.; Foster, T.J. Interactions between cellulose ethers and a bile salt in the control of lipid digestion of lipid-based systems. *Carbohydr. Polym.* **2014**, *113*, 53–61. [CrossRef] [PubMed]

Article

Comparative Evaluation on Impacts of Fibronectin, Heparin–Chitosan, and Albumin Coating of Bacterial Nanocellulose Small-Diameter Vascular Grafts on Endothelialization In Vitro

Max Wacker [1,*], Jan Riedel [1], Heike Walles [2], Maximilian Scherner [1], George Awad [1], Sam Varghese [1], Sebastian Schürlein [3], Bernd Garke [4], Priya Veluswamy [1], Jens Wippermann [1] and Jörn Hülsmann [1]

1 Department of Cardiothoracic Surgery, University Hospital Magdeburg, 39112 Magdeburg, Germany; jan.n.riedel@icloud.com (J.R.); maximilian.scherner@med.ovgu.de (M.S.); george.awad@med.ovgu.de (G.A.); sam.varghese@med.ovgu.de (S.V.); priya.veluswamy@med.ovgu.de (P.V.); jens.wippermann@med.ovgu.de (J.W.); joern.huelsmann@med.ovgu.de (J.H.)
2 Core Facility Tissue Engineering, Otto-Von-Guericke University Magdeburg, 39106 Magdeburg, Germany; heike.walles@ovgu.de
3 Department Tissue Engineering and Regenerative Medicine (TERM), University Hospital Würzburg, 97070 Würzburg, Germany; sebastian.schuerlein@gmail.com
4 Institute of Experimental Physics, Otto-Von-Guericke University Magdeburg, 39106 Magdeburg, Germany; bernd.garke@ovgu.de
* Correspondence: max.wacker@med.ovgu.de; Tel.: +49-391-67-14102

Citation: Wacker, M.; Riedel, J.; Walles, H.; Scherner, M.; Awad, G.; Varghese, S.; Schürlein, S.; Garke, B.; Veluswamy, P.; Wippermann, J.; et al. Comparative Evaluation on Impacts of Fibronectin, Heparin–Chitosan, and Albumin Coating of Bacterial Nanocellulose Small-Diameter Vascular Grafts on Endothelialization In Vitro. *Nanomaterials* **2021**, *11*, 1952. https://doi.org/10.3390/nano11081952

Academic Editors: Rajesh Sunasee and Karina Ckless

Received: 29 June 2021
Accepted: 25 July 2021
Published: 29 July 2021

Publisher's Note: MDPI stays neutral with regard to jurisdictional claims in published maps and institutional affiliations.

Abstract: In this study, we contrast the impacts of surface coating bacterial nanocellulose small-diameter vascular grafts (BNC-SDVGs) with human albumin, fibronectin, or heparin–chitosan upon endothelialization with human saphenous vein endothelial cells (VEC) or endothelial progenitor cells (EPC) in vitro. In one scenario, coated grafts were cut into 2D circular patches for static colonization of a defined inner surface area; in another scenario, they were mounted on a customized bioreactor and subsequently perfused for cell seeding. We evaluated the colonization by emerging metabolic activity and the preservation of endothelial functionality by water soluble tetrazolium salts (WST-1), acetylated low-density lipoprotein (AcLDL) uptake assays, and immune fluorescence staining. Uncoated BNC scaffolds served as controls. The fibronectin coating significantly promoted adhesion and growth of VECs and EPCs, while albumin only promoted adhesion of VECs, but here, the cells were functionally impaired as indicated by missing AcLDL uptake. The heparin–chitosan coating led to significantly improved adhesion of EPCs, but not VECs. In summary, both fibronectin and heparin–chitosan coatings could beneficially impact the endothelialization of BNC-SDVGs and might therefore represent promising approaches to help improve the longevity and reduce the thrombogenicity of BNC-SDVGs in the future.

Keywords: bacterial nanocellulose; small-diameter vascular grafts; endothelialization; tissue engineering; bioreactor

1. Introduction

Coronary artery bypass grafting (CABG) is the most frequently performed heart surgeries in the western world. In Germany and the USA alone, more than 450,000 CABG operations are conducted every year, using small caliber vascular grafts with an inner diameter of 6 mm or below to redirect the blood distal to the blockage [1,2]. Autologous vessels—especially the internal thoracic artery, the radial artery, or the saphenous vein—are considered the gold standard for graft material in CABG surgery, as they represent the best compromise between availability and long-term patency rate, which is mainly dependent on the biological function of the grafts, e.g., the endothelial integrity and mechanical characteristics of the vessel wall [3,4]. While the optimal bypass graft has not yet been found, access to autologous vessels that are expected to have a long-term patency of over 90%, such as the internal thoracic artery, is often limited due to vascular diseases, pre-surgery, or increased risk of wound infections [5–7], and alternatives, such as the saphenous vein, are expected to occlude in over 50% of the cases after 10 years [8]. Therefore, alternatives are warranted, and today, tissue-engineered vascular grafts (TEVG) are thought to be a promising approach in the development of synthetic grafts mimicking the biological functionality of autologous vessels with unlimited availability.

Despite many different approaches that have been examined to date, none have been able to show satisfactory long-term patency in application-oriented experimental setups [9]. Common causes of low patency rates, such as intimal hyperplasia and thrombus formation, could be addressed with a functional endothelium [10,11]. However, the time-consuming fabrication of in vitro endothelialized grafts does not appear to be appropriate, as current guidelines recommend revascularization within a few days to a few weeks following diagnosis [12]. Therefore, acellular grafts for off-the-shelf use that become endothelialized in situ after implantation are warranted. Among the variety of different approaches in the development of small-diameter TEVGs, acellular scaffolds made of synthetic polymers are of particular interest, because they are reproducible, possess satisfying mechanical properties, and are capable of being stored for longer periods without the need for long-term precultivation and cell seeding [13].

We developed an acellular TEVG composed of bacterial nanocellulose (BNC) with a small inner diameter (below 5.00 mm). The BNC grafts have shown satisfactory short-term results, which were characterized by good surgical manageability and burst pressure in the physiological environment. These grafts were also successfully surface modified to reduce the thrombogenicity, but the long-term results are still limited as reflected by low patency rates and lack of sufficient endothelialization [14–17].

The in vivo endothelialization of acellular scaffolds after implantation works by two different mechanisms: (i) continuous ingrowth of vascular endothelial cells (VECs) from the anastomotic region and (ii) attachment of circulating endothelial progenitor cells (EPCs) directly from the blood [18]. The homing and adhesion of VECs and EPCs is mainly dependent on expression of proteins on the respective cell surface; in fact, surface coating with substances that resemble the function of these proteins has shown promising results with regard to adhesion of endothelial cells [18,19].

With regard to BNC grafts, coatings with fibronectin, heparin–chitosan, or albumin are interesting approaches for future surface modification of BNC grafts. Fibronectin, of note, is a protein of the extracellular matrix modulating the adhesion and proliferation of endothelial cells in vivo [20]. It has been shown that fibronectin coating led to significant improvements with endothelial cell expansion on the BNC surface [21]. Crosslinking of BNC with heparin led to increased endothelialization and reduced activation of the coagulation cascades [22,23], and heparin coating could be of further interest with respect to intimal hyperplasia, as it is reported to inhibit the growth of fibroblasts [24]. This is particularly appealing, as fibroblast growth can result in intimal hyperplasia and thus late occlusion of the synthetic grafts. Chitosan, a polysaccharide derived from the chitin shell of crustaceans with antibacterial and anticoagulative properties, has recently shown excellent results with regard to endothelial cell adhesion and proliferation when used as a component of small-diameter vascular grafts [25,26]. A combination of heparin and chitosan has already been introduced as a coating for BNC based grafts but was only examined with non-human endothelial cells [22]. This encouraged us to test this combination in a more clinical setup using human VECs. Additionally, albumin also seems to be a promising coating to induce EPC homing for an enhanced in situ self-endothelialization on vascular devices [27].

However, more information to verify these beneficial impacts for clinically relevant cells by proof of principle studies in a dynamic environment is still warranted to help the design and further development of TEVGs, as many studies investigating the effect on the above mentioned coatings on endothelialization have been conducted using non-human cells or human umbilical vein endothelial cells [21,23,28,29], which differ significantly in function and receptor expression from human saphenous vein endothelial cells or EPCs [30,31], and which therefore relativize the translational approach of these studies. This translational approach is further limited when in vitro experiments do not include medium perfusion [21,22,32–34], because the activity of integrins that mediate cell adhesion is dependent on shear stress and is regulated accordingly [35]. With this comparative study, we report the endothelialization of small-diameter vascular grafts made of BNC which were either coated with fibronectin, heparin–chitosan, or albumin in a clinically relevant in vitro setting and tried to simulate the natural and necessary stimulation by wall shear stress in a bioreactor.

Upon cell seeding with human primary EPCs or human saphenous vein endothelial cells, these grafts were cultivated under both static and dynamic conditions, using an in vitro approach and a bioreactor perfusion system. Besides proliferation assays, the endothelialized grafts were analyzed histologically with regard to the extent of endothelialization and endothelial cell characterization, including their functionality. Thus, using the current state-of-the-art equipment, we hope to better screen for the potential for translational applications.

2. Materials and Methods

All procedures were in accordance with the ethical standards of the responsible committee on human experimentation of the University Hospital of Magdeburg (application number 81/18) and with the Helsinki Declaration of 1975, as revised in 2000. Informed consent was obtained from all participants for being included in the study.

2.1. BNC Graft Production

The BNC grafts with an inner diameter of 5.0 mm were produced in a standardized fashion as described before [36–38]. Briefly, in a customized bioreactor, cylindrical templates made of bamboo with a diameter of 5.0 mm were moved periodically between air space and a reservoir, which was filled with liquid culture medium and bacteria of the genus Komagataeibacter xylinus (German Collection of Microorganisms and Cell Cultures GmbH, DSM 32384). During dipping, the template was loaded with culture medium and bacteria, and after leaving the liquid, BNC formation took place on the template surface. After the cultivation time, the templates were removed from the BNC, leaving a vascular graft with an inner diameter that corresponded to the bamboo template. The BNC grafts then underwent several post-processing steps, including washing in ultrapure water and boiling in sodium hydroxide solution, followed by autoclaving.

2.2. Coating Procedures

To investigate the effect of surface coating on the in vitro endothelialization, three different coatings (human fibronectin, heparin–chitosan, and human albumin) were examined and compared to uncoated BNC grafts. Thus, the different coated BNC grafts are referred to as fibronectin, heparin, albumin, or uncoated group. Static culture was performed in 48-well plates, where round patches of 10 mm diameter were punched out from the tubular, uncoated grafts with a medical biopsy punch and were subsequently coated. For dynamic culture, the coating was performed directly on the tubular graft. All coating procedures for the patches and the tubular BNC scaffolds in the heparin group were performed in a beaker glass. For fibronectin and albumin coating of the tubular BNC scaffolds, autoclaved BNC grafts were mounted to the custom-made 3D printed bioreactors and fixed with non-absorbable surgical sutures under sterile conditions as shown in Figure 1. The coating procedure for the grafts was then performed by adding the coating solutions with syringes directly to the mounted grafts through the luer-lock adapters.

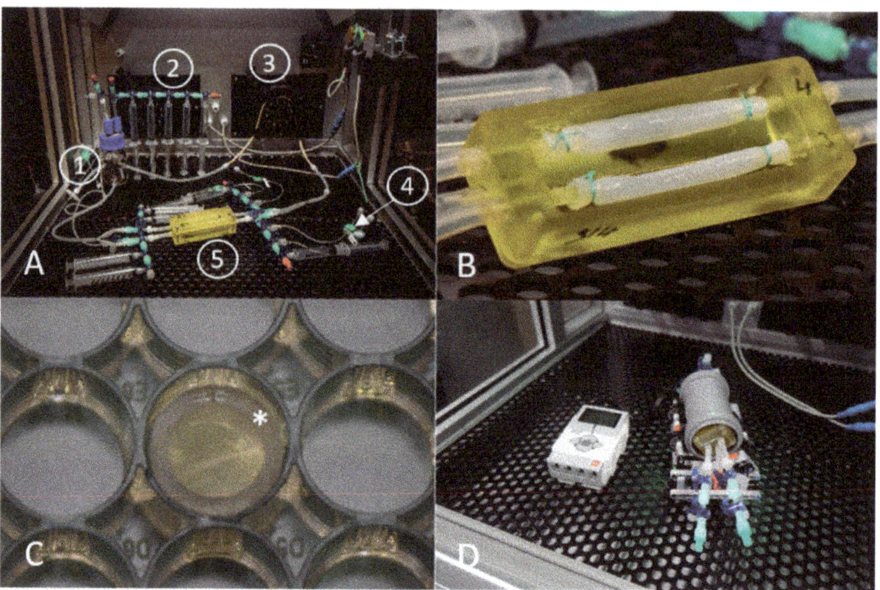

Figure 1. Laboratory setup for dynamic and static endothelialization experiments: (**A**) Perfusion system which was set up in an incubation chamber with controlled atmosphere (5% CO_2). (1) Medium reservoir, (2) Syringes for media sampling, (3) Roller pump, (4) 0.2 µm sterile filter for adding WST-1 cell proliferation assay reagent, (5) 3D printed reactor containing the BNC grafts. (**B**) Enlarged view of the BNC grafts mounted on the 3D printed bioreactor, secured with surgical sutures. (**C**) A piece auf autoclaved PVC tube (*) was added on top of the patch and placed in a 48-well plate to avoid cells running off the BNC patch during the first 24 h of static cultivation. (**D**) Perfusion chamber mounted to a self-made rotation unit. Left: Control unit. Right: The bioreactor was positioned inside a commercially available piece of drain pipe and rotated on the rotation unit to promote cell distribution over the whole BNC graft surface.

2.2.1. Heparin–Chitosan Coating

The heparin–chitosan coating on BNC was done in a modified way than originally described in 2017 by Li et al. [22]. Briefly, the BNC constructs were immersed in an aqueous chitosan solution containing 1% (*v*/*v*) acetic acid (Carl Roth, Karlsruhe, Germany), 1% (*v*/*v*) glycerin (Sigma-Aldrich, St. Louis, MI, USA) and 1 mg/mL chitosan (molecular weight 600,000 to 800,000 Da, ≥90% deacetylated; VWR Chemicals, Radnor, PA, USA) for 24 h under agitation at room temperature (RT), followed by fixation in 0.1 M NaOH aqueous solution (Sigma-Aldrich, St. Louis, MI, USA) for 24 h without agitation. After extensive washing with 1% (*v*/*v*) penicillin/streptomycin (pen-strep) in distilled water (ThermoFisher Scientific, Waltham, MA, USA), the BNC constructs were directly transferred into a glass beaker containing 0.05 M MES buffer (Sigma-Aldrich, St. Louis, MI, USA) for 2 h. Then, the BNC scaffolds were coated with heparin by immersion in an EDC/NHS crosslinking solution for the formation of amide and ester bonds, prepared with 0.05 M MES buffer, containing 1 mg/mL heparin, 0.25 M NaCl, 0.03 M NHS, and 0.06 M EDC (all purchased from Sigma-Aldrich, St. Louis, MI, USA) for 24 h in a water bath shaker at 30 °C. After extensive washing with pen-strep solution for 2 h, the tubes were stored in pen-strep solution, sterilized by gamma irradiation with 13 Gray, and stored at 4 °C.

2.2.2. Albumin and Fibronectin Coating

The coating procedure was modified from the procedure published by Kuzmenko et al. [21]. The 1-cyano-4-dimethylamino pyridinium tetrafluoroborate (CDAP), acetonitrile, triethylamine sodium acetate, human fibronectin, and human albumin were purchased from Sigma-Aldrich (St. Louis, MI, USA). First, the BNC surface was activated with CDAP as the respective crosslinker, generating a cyanate ester derivate. For this, a CDAP crosslink solution was prepared by dissolving 50 mg CDAP in 1 mL acetonitrile. For activation, the required amount of CDAP per cm^2 patch surface for the inner graft surface was ensured by setting the concentration to 50 $\mu mol/cm^2$ in an excess of volume, ensuring that the patches were completely immersed or the tubes were completely filled with CDAP crosslink solution. After adding the CDAP crosslink solution to the BNC grafts or patches, they were incubated for 30 s at room temperature. Then, the same volume of 0.2 M triethylamine (TEA) solution prepared in distilled water was added and incubated for 2 min, followed by discarding the CDAP/TEA solution. After rinsing the BNC patches/grafts with 0.1 M sodium acetate buffer (pH 3.0), prepared in distilled water, the BNC patches/grafts were extensively rinsed with 1% pen-strep solution until the pH was neutralized.

For subsequent bioconjugation with the respective protein by an isourea bond to the amino acids of the protein, an aqueous solution containing 100 μg/mL human fibronectin or albumin was added to the CDAP activated BNC patches/grafts, ensuring that the patches were completely covered or no air remained in the tubular grafts. After incubation for 12 h at 37 °C, the patches and grafts were rinsed with phosphate buffered saline solution containing 1% (v/v) pen-strep, gamma irradiated (13 Gray for patches and 25 Gray for tubular grafts), and stored at 4 °C.

2.3. *Coating Analyses for Quality Control*

2.3.1. XPS Analyses

X-ray photoelectron spectroscopy (XPS) was used to analyze the surface chemistry of the coated BNC grafts [39]. Therefore, rectangular samples (0.5 × 0.5 cm) were cut from coated BNC grafts and freeze dried. Then, the samples were analyzed by XPS using a PHI 5600 ESCA System (Physical Electronics Inc., Division of ULVAC-PHI, 187,25 Lake Drive East in Chanhassen, MN 55317, USA), equipped with a Spherical Capacitor electron energy Analyzer (energy resolution 25 meV, PHI 5600 ESCA System) and a Dual Anode X-ray Source operating with Mg K-alpha at 14 kV and 400 W. The analyzed lateral area had a diameter of 800 μm, using a detection angle (Source to SCA) of 54°. To define the atomic concentration of the analyzed elements, multiplex spectra were measured for each sample.

2.3.2. Heparin Detection

To verify sufficient heparin presence at the graft surface, we used a toluidine blue staining modified from the one described by Hinrichs et al. [40]. Samples of coated and uncoated BNC grafts were stained in a 0.04% toluidine blue (Sigma-Aldrich, St. Louis, MI, USA) staining solution prepared in 0.01 M HCL containing 2 mg/mL NaCl for 12 h, followed by extensive washing with distilled water and incubation for 24 h in distilled water before images were obtained.

2.3.3. Albumin and Fibronectin Detection

To validate the presence of the respective functional proteins at the luminal surface after coating, grafts were sampled and subjected to immunofluorescent evaluation. Therefore, 0.5 cm long sections from exemplary tubular grafts coated with albumin or fibronectin were frozen in compound (Tissue-Tek O.C.T. compound, Sakura Finetek Europe B.V., Alphen aan den Rijn, NL) and thereafter sectioned using a Leica CM 1950 cryostat (Leica Biosystems, Nussloch, Germany). Recombinant monoclonal rabbit anti-human serum albumin antibody (MA5-29022, ThermoFisher Scientific, Waltham, MA, USA) and recombinant monoclonal rabbit anti-human fibronectin antibody (MA5-32509, ThermoFisher Scientific, Waltham, MA, USA) were used as the primary antibody and donkey anti-rabbit Alexa Fluro 488 (Jackson ImmunoResearch, West Grove, PA, USA) as the secondary antibody. Upon fixation in 4% paraformaldehyde (PFA), the graft slices were blocked with 3% standard donkey serum for 30 min and then incubated with the primary antibodies at a dilution of 1:50 in 3% donkey serum overnight. After washing, the fluorescent dye-conjugated secondary antibody at a dilution of 1:500 in 3% donkey serum was incubated on the slice in the dark for 1 h. After mounting, the slices were analyzed using the EVOS Auto 2 (ThermoFisher Scientific, Waltham, MA, USA).

2.4. Cell Isolation and Cell Culture

2.4.1. Human Saphenous Vein Endothelial Cells (VECs)

The isolation of endothelial cells from human saphenous veins for use in vascular prostheses seeding experiments is a standard method and was described in 1984 by Watkins et al. [41]. For cell isolation, the vein sections obtained from patients undergoing coronary artery bypass surgery were flushed with PBS to remove blood and thereafter filled with 0.4% Collagenase A solution (Sigma-Aldrich, St. Louis, MI, USA) and incubated at 37 °C for 30 min. By rinsing with PBS, the cell suspension was collected and plated as passage 0 on 0.1% gelatin-coated cell culture flasks using a cell specific growth medium (Endothelial Cell Growth Medium C22110, Promocell, Heidelberg, Germany) containing 1% pen-strep solution and 10% fetal calf serum (FCS) with a density of 0.6–1.0 $\times$ 10^5 cells/cm^2. For long-term stand cell culture, the cells were passaged at 90% confluence using 0.05% trypsin/EDTA (ThermoFisher Scientific, Waltham, MA, USA). Also, for the standard growth medium, the FCS content was reduced to 5% for passage 1 and 2% for later passages. For seeding experiments, endothelial cell cultures were used until passage 5 as a maximum. Cells were stored at −150 °C (50% FCS, 10% DMSO).

2.4.2. Human Endothelial Progenitor Cells (EPCs)

Human endothelial progenitor cells were isolated from peripheral blood obtained from healthy volunteers. The procedure was carried out as described by Ormiston et al. [42]. Briefly, 60 mL of fresh blood was drawn in sodium citrate tubes (Vacutainer 367704, Becton Dickinson [BD] GmbH, Heidelberg, Germany), and peripheral blood mononuclear cells (PBMCs) were obtained by density gradient centrifugation (Ficoll-Paque Plus, GE Healthcare, IL, USA). After washing with PBS, the PBMCs were resuspended in endothelial cell medium (Endothelial Growth Medium-2MV, CC-3202, Lonza, Basel, Switzerland) with 10% FCS content and plated in T75 flasks coated with 5 µg/cm^2 collagen (Type 1 Collagen, derived from rat tail, 35-4236; BD Biosciences, Heidelberg, Germany).

According to the protocol, the cells were passaged for the first time after initial outgrowth colonies reached a size of approximately 1000–2000 cells/colony, counted on microscopy images. Deviating from the protocol of Ormiston et al., TrypLE (ThermoFisher Scientific, Waltham, MA, USA) was used for cell dissociation. The cells were cultured in uncoated TPP tissue culture flasks (Techno Plastic Products AG, Trasadingen, Switzerland) after passage 5 with an FCS content of 5% for long-term cell culture. The cells were passaged at 90% confluence, cultivated to a maximum of nine passages and stored at −150 °C (50% FCS, 10% DMSO).

2.5. Cell Characterization

A detailed description of endothelial cell characterization is described in the Supplementary Material (Figures S1 and S2).

2.5.1. Phase Contrast Microscopy

For morphologic characterization, cells were observed daily with a conventional phase contrast microscope during culture (EVOS XL Core Imaging System, ThermoFisher Scientific, Waltham, MA, USA). In the case of apparent fibroblast contamination, those cultures were not used for experiments.

2.5.2. AcLDL Assay

To exclude potential degeneration of the endothelial cells, we actively tested the endothelial functionality by the ability for the uptake of acetylated low density lipoproteins by the scavenger cell pathway using a commercial assay (Alexa Fluor 488 AcLDL assay, ThermoFisher Scientific, Waltham, MA, USA) [43–45]. For AcLDL assay, the endothelial cells were seeded in 24-well plates and cultivated until 90% confluence. The assay was performed as recommended by the manufacturer. Briefly, the cells were washed two times in a 1% bovine serum albumin (BSA)/PBS$^+$ solution for 5 min, followed by adding 400 µL of cell culture medium containing 0.0.125% AcLDL assay solution and incubation at 37 °C for 3.5 h. After adding one drop of NucBlue live staining solution for Deoxyribonucleic acid (DNA) staining, (ThermoFisher Scientific, Waltham, MA, USA), the cells were incubated for another 30 min in the incubator. After washing another three times, images were directly obtained with a conventional fluorescent microscope (EVOS FL Auto 2 cell imaging system, ThermoFisher Scientific, Waltham, MA, USA).

2.5.3. Flow Cytometry

The isolated endothelial cells were characterized using flow cytometry (FC) for the expression of the endothelial cell markers von Willebrand Factor, CD 144, CD31, and CD34. A detailed description of the methodology and the results is given in the supplementary files.

2.6. Control Based Construct Culture

To get a general impression of the cell growth characteristics on the corresponding BNC surfaces, we chose a static model of cultivation in 48-well plates, and to achieve further understanding of the cell adhesion property in a so-called physiological environment, we cultivated the cells under media perfusion. For both dynamic and static conditions, a control based approach was chosen, where cells were seeded in the same density on 0.1% gelatin-coated well plates and used for normalization of metabolic activity, where values from control cells were set to 100% as optimal cell adherence to the cell culture flask bottom and respective proliferation was considered. To determine optimal cell densities for seeding, previous examinations were conducted where cells were seeded on uncoated BNC patches and gelatin-coated control well plates. After static incubation for 24 h and DNA staining, the cell distribution was examined with immunofluorescence microscopy, determining the lowest cell concentration that avoids overly dense growth in the control wells but still reaches a seeding efficacy of over 5%, as reported in the literature for endothelialization-based cell seeding experiments [46]. Based on these results, the seeding density was set to 5×10^4 cells per cm^2 surface area. A concentration of 10% FCS in the respective cell culture medium was chosen to provide optimal nutrients to the endothelial cells. 2×10^6 cells from each experiment were frozen and stored at −150 °C for subsequent cell characterization (50% FCS, 10% Dimethylsulfoxid (DMSO)).

2.6.1. Static Culture Experiments

The coated and uncoated patches were placed in a 48-well plate with the luminal side facing upwards. The position inside the well was fixed by placing a piece of autoclaved polyvinylchloride (PVC) tube (inner diameter of 8 mm, wall thickness of 1 mm) inside the well (Figure 1), thereby ensuring that cells did not run off the BNC patch surface. Either VECs (n = 10 for each coating) or EPCs (n = 5 for each coating) were thawed and cultured for 4 days in T75 flasks on 0.1% gelatin coating. After harvesting, the patches were subjected to passive seeding and the cells were allowed to settle. After 24 h cultivation for cell settling and adhesion, the PVC tubes were removed and the patches were placed into new wells for subsequent cultivation for 96 h. The media was changed every day.

2.6.2. Perfusion Culture Experiments

Bioreactor Perfusion System

To study the behavior of the seeded endothelial cells on the grafts under perfusion, we used customized 3D printed bioreactors and a perfusion platform that was described by Schuerlein et al. [47], shown in Figure 1A. The 3D printed bioreactors allowed us to perfuse two BNC tubes connected to one pump system simultaneously (Figure 1B). All parts of the reactor were autoclavable and the BNC grafts were attached to the connectors with surgical sutures. To exclude possible bias from different lengths of the silicone tubes, the volume inside the perfusion path between the two stop cocks upstream and downstream to the BNC tube was determined individually for each perfusion slot. The perfusion system further consisted of silicon tubes and a reservoir bottle as shown in Figure 1A. After autoclaving, the system was assembled under the cell culture bench, and syringes for medium sampling as well as 0.2 µm sterile filters (Intrapur Neonat infusion filter, B. Braun Melsungen AG, Germany) for the cell proliferation assay were attached before the closed perfusion system was mounted on the perfusion platform. The perfusion platform consisted of an incubator with a controlled atmosphere (5% CO_2) and temperature (37 °C) and a roller pump for medium perfusion.

Seeding Technique

The respective endothelial cells were thawed and cultured 4 days in advance in T75 flasks with 0.1% gelatin coating for VECs or no coating for EPCs and 10% FCS. On the first day of the experiment, the cells were harvested after setting the cell concentration; the suspension was taken up in a 10 mL syringe. Then the cell suspension was manually perfused through the perfusion path that integrates the graft in between two three-way stopcocks, holding the bioreactor in an upright position, until the whole graft was filled with the cell suspension, and the connectors were closed with sterile caps. Cell settling and adhesion were supported by incubation on a custom-made roller shaker (Figure 1D) at a speed of 0.05 rpm at 37 °C for 30 min, followed by static incubation for 3.5 h at 37 °C. After incubation, the bioreactors were connected to the tubing system as described before and mounted on the perfusion platform. The perfusion pump was set to pulsatile flow and a medium flow rate of 1.5 mL/min. The perfusion system was filled with a total of 50 mL of medium and samples were taken every day for analytical evaluation. After 96 h, the perfusion was stopped and the BNC grafts were removed, divided into equal parts, and stored in PBS for subsequent analyses. Cells seeded on 0.1% gelatin-coated (VEC) or uncoated (EPC) 12-well plates with the same cell number were used as controls. For each coating group, $n = 4$ BNC grafts were cultured and analyzed.

2.7. Metabolic Activity (WST-1 Assay) of Cells on BNC Constructs, Normalization

A WST-1 ((4-(3-4-iodophenyl)-2-(4-nitrophenyl)-2H-5-tetrazolio)-1,3-benzenedisulfonate) proliferation assay (Sigma-Aldrich, St. Louis, MI, USA) was used to determine the overall cell viability and proliferation of the cells. For the assay, a working solution was prepared by diluting the WST-1 reagent in the respective cell culture medium to a final concentration (v/v) of 12.5% in growth medium. The WST-1 cell proliferation assay was conducted under both static and dynamic conditions and for both cell lines, VEC and EPC.

2.7.1. Static Cultures

The WST-1 cell proliferation assay was performed after 24 h and 96 h of cultivation. 500 µL of the WST-1 working solution was added to the respective wells after washing with PBS for an incubation time of 4 h at 37 °C.

2.7.2. Perfusion Cultures

Consistent with the static cultivations, WST-1 proliferation assay was performed after 24 h and 96. The perfusion was stopped and 3 mL of WST-1 working solution was added to the lumen of each BNC graft individually via the connected 0.2 µm filters (Intrapur Neonat infusion filter, B. Braun Melsungen AG, Germany) to avoid contamination. After 4 h of incubation, the WST-1 working solution was collected and the perfusion was re-established. In parallel, the positive controls in the 12-well plate were incubated in 437 µL of WST-1 working solution, which corresponded to the same amount of WST-1 volume per seeded cells per surface area.

The extinction of the WST-1 working solution was determined for each sample with a multiplate reader (Infinite 200 pro M Plex, Tecan Trading AG, Switzerland) at 440 nm (reference wavelength 600 nm) and blanked to working solution. All samples were measured as duplicates. For dynamic conditions, the measured absorbance had to be multiplied with a dilution factor, depending on the luminal volume in the perfusion path. The values for WST-1 proliferation index obtained after 24 h and 96 h were normalized to controls as described before.

2.8. AcLDL Assay on Seeded BNC Cell Constructs

The AcLDL assay was performed as described above with light modifications. After 96 h of cultivation, the seeded BNC constructs (patches or sections from tubular BNC grafts) were immersed in 2 mL Eppendorf tubes filled with PBS for washing. After 5 min of incubation, the samples were transferred to a new micro-reaction tube with fresh PBS. The washing step was repeated three times. Afterwards, the AcLDL assay was performed as described before with an excess of medium containing 0.0125% AcLDL assay solution. Following the incubation time, the samples were washed again in PBS and images were acquired with a fluorescence microscope (EVOS FL Auto 2 cell imaging system, ThermoFisher Scientific, Waltham, MA, USA). Subsequently, the samples were stored in PBS at 4 °C until they were additionally stained with rhodamine-conjugated phalloidin dye as described above. Finally, further images were taken with a Leica confocal imaging system (Leica TCS SP8, Leica Microsystems GmbH, Wetzlar, Germany).

2.9. Cryosectioning and Immunofluorescence of BNC Constructs

Acridine Orange staining, a fluorescent dye that reports both cytoplasmic RNA and DNA in the nuclei, was used to get a visual impression of cell survival, cell morphology, and its distribution on the BNC surface [48,49]. To determine the shape of the cytoskeleton of the endothelial cells, we additionally stained for F-actin, which is a commonly accepted method for morphological studies on endothelial cells [50,51]. Furthermore, we complemented a CD-31 staining to detect preserved endothelial character of the cells [50]. For cultivated constructs, cell-specific markers were evaluated after fixation in 4% PFA solution for 5 min before staining. Acridine orange staining was achieved by incubating the whole sample in an acridine orange staining solution for 45 min, prepared from acridine orange dye (Sigma-Aldrich, St. Louis, MI, USA) diluted 1000-fold in PBS. Samples were subjected to F-actin staining, and subsequently to the AcLDL uptake assay. Therefore, we used rhodamine-conjugated phalloidin dye (Abcam, Cambridge, UK) diluted 1:1000 stock solution in PBS containing 1% BSA. Nuclei were counterstained with NucBlue. CD31 staining was performed utilizing CD31/PECAM-1 mouse anti-human antibody (BBA7, R&D Systems, Minneapolis, MN, USA) in a concentration of 8 μg/mL as the primary antibody and donkey anti-rabbit Alexa Fluro 488 (Jackson ImmunoResearch, West Grove, PA, USA) at the dilution of 1:500 as the secondary antibody. Both antibodies were diluted in PBS− containing 3% donkey serum. After overnight incubation at 4 °C with the primary antibody, samples were washed with PBS−. For secondary antibody incubation, the samples were incubated for another hour at room temperature in a staining solution consisting of the secondary antibody, 1:1000 rhodamine-conjugated phalloidin dye, and NucBlue DNA staining. Finally, the samples were washed again and images were obtained with a Leica confocal imaging system (SP8) or with a conventional immunofluorescence microscope (EVOS FL Auto 2 cell imaging system, ThermoFisher Scientific, Waltham, MA, USA).

2.10. Data Analysis and Image Processing

Due to the low n-numbers, non-normally distributed errors can be expected for the raised data, and thus a non-parametric Kruskal–Wallis test with Dunn's post-test was performed using GraphPad Prism version 8 for Windows, (GraphPad Software, La Jolla, CA, USA). The level of significance was set to $p < 0.05$.

Fiji 64-bit for Windows (Version 1.53c) [52] was used for image processing if not otherwise stated. For stitching the whole mount images of acridine orange-stained samples, the plugin "Stitching" of Fiji was used, based on the publication from Preibisch et al. [53], following the directions as given by the plugin description.

3. Results

3.1. Comparative Coating Efficiencies of Fibronectin-, Heparin–Chitosan- and Albumin-Coated BNC Grafts

In immunofluorescence images, the coated BNC grafts showed one homogenous layer of albumin and fibronectin at the luminal side of the grafts; in particular, no discontinuation of the layer was found (Figure 2A,B). Accordingly, the concentration of nitrogen at the luminal surface was clearly enhanced only for protein-coated groups, showing a concentration of 5.6% for albumin- and 6.7% for fibronectin-coated grafts on the luminal surface, as shown by XPS analyses (Figure 2D). For uncoated grafts, the concentration of nitrogen was 0%. The heparin-coated grafts stained with toluidine blue exhibited a deep blue staining covering the whole surface, compared to uncoated grafts (Figure 2C), and the detection of sulphur (1.1% vs. 0% for uncoated or albumin-/fibronectin-coated grafts) on the luminal side by XPS analyses indicated the presence of the highly sulphated glycosaminoglycan heparin. Data on the stability of the albumin and fibronectin coating are included as Table S2 in the supplementary files.

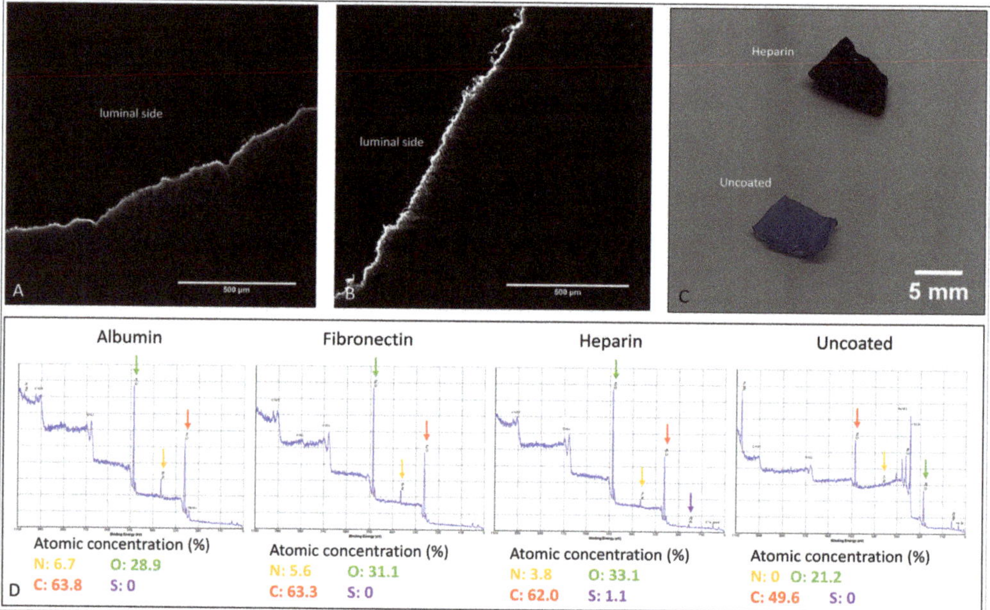

Figure 2. Coating analyses of exemplary coated BNC grafts: (**A**) Immunofluorescence staining of a cross-sectioned BNC graft against albumin. (**B**) Immunofluorescence staining of a cross-sectioned BNC graft against fibronectin. (**C**) Comparison of toluidine blue staining for heparin-coated and uncoated BNC grafts. (**D**) Results of XPS analyses of the coated BNC grafts. The colored arrows indicate the respective peaks for N (nitrogen), O (oxygen), C (carbon), and S (Sulphur).

3.2. Progress of Metabolic Activity

The results are summarized in Figure 3 and Table S1. In general, increased metabolic activity (as indicated by measured extinction of the WST-1 metabolite after 24 h of incubation) was considered to be increased vitalization (Figure 3). Interestingly, values clearly varied for different cell types. For static culture after 24 h, only the fibronectin group showed a recognizable vitalization by VEC cells, which revealed significantly higher metabolic activity compared to heparin-coated (18.93 ± 7.40 vs. 2.00 ± 3.12%, $p = 0.0001$) and uncoated grafts (1.07 ± 2.28%, $p < 0.0001$). After 96 h, this had not significantly changed, then showing 27.50 ± 9.74% for the fibronectin group, which was significantly higher than the heparin group (1.40 ± 4.55%, $p < 0.0001$) and uncoated group (1.70 ± 4.19%, $p < 0.0001$). The albumin group showed 6.00 ± 4.50% after 24 h and 9.70 ± 8.67% after

96 h, not reaching the level of statistical significance compared to the other groups. For dynamic cultivations, the vitalization of all groups by VEC cells had clearly increased to 51.00 ± 14.88% for albumin, 59.25 ± 23.92% for fibronectin, 32.00 ± 10.42% for heparin, and 32.50 ± 27.45% for the uncoated group, not showing statistically significant differences. After 96 h by trend, this had slightly decreased to 33.25 ± 10.44% for albumin, 44.25 ± 14.55% for fibronectin, 29.75 ± 13.57% for heparin, and 34.00 ± 31.78% for the uncoated group.

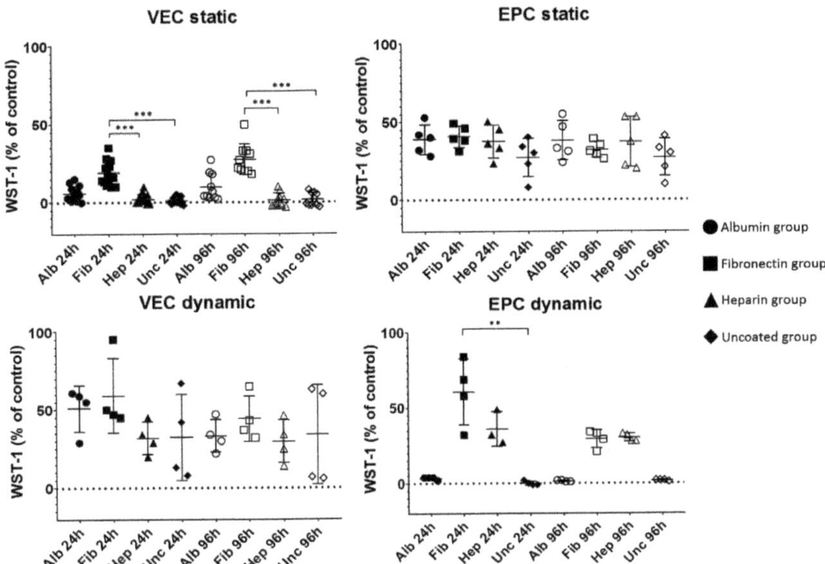

Figure 3. Results of WST-1 proliferation assay performed on the cell seeded BNC grafts under both static and dynamic conditions for vascular endothelial cells (VEC) and endothelial progenitor cells (EPC). The WST-1 values are normalized to control cells. $n = 10$ (VEC static), $n = 5$ (EPC static), $n = 4$ (VEC and EPC dynamic except EPC Hep 24 h ($n = 3$)). Alb—albumin, Fib—fibronectin, Hep—heparin, Unc—uncoated BNC grafts. ** $p < 0.01$, *** $p < 0.001$.

For EPC, similar values for all coatings could be observed after static culture for 24 h, where values ranged between 27.00 ± 12.29% for uncoated and 40.60 ± 7.09% the fibronectin, without reaching the level of statistical significance. After 96 h of culture, this had not changed, still showing similar values for all coatings. Dynamic culture led to clearly higher values after 24 h, but only for fibronectin (60.75 ± 21.93%) and heparin (36.00 ± 11.53%), while the others remained close to zero. After 96 h, both groups decreased to 29.25 ± 5.91% for fibronectin and 30.25 ± 2.63% for heparin, while the others still remained close to zero.

3.3. Histological Analyses

3.3.1. Graft Colonization under Static Conditions: Acridine Orange Staining

Figure 4 shows whole mount images after cultivation for 96 h under static conditions, stained with acridine orange. The round image sections correspond to the total area of BNC patches seeded with endothelial cells and cultured for 96 h as described earlier.

For VECs, only fibronectin-coated patches showed multiple large and partially coherent colonized areas, but also nonpopulated spots and areas, especially close to the edge where the formerly attached pieces of PVC tube were imposed. All other coatings only showed smaller cell islets or singular cells. In contrast, for EPC, densely populated areas of EPCs were found for all coatings and also uncoated BNC patches after seeding the cells under static conditions.

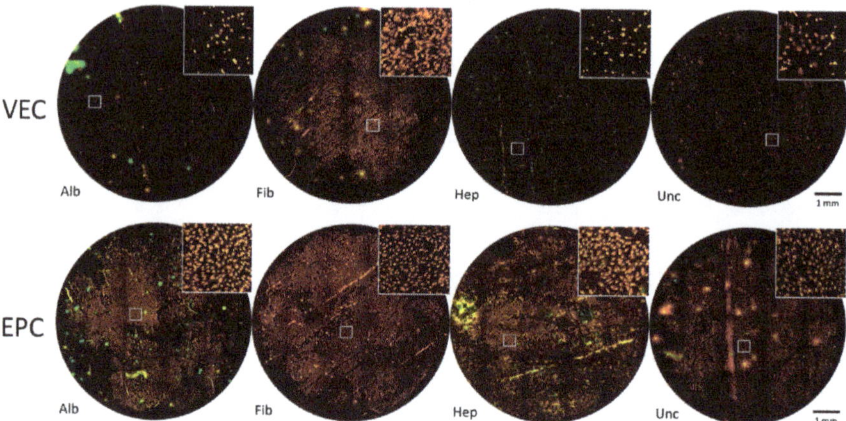

Figure 4. Stitched whole-mount microscopy images of acridine orange staining of BNC grafts seeded with vascular endothelial cells (VECs, upper row) and endothelial progenitor cells (EPCs, bottom row) under static conditions. The region of interest outlined in white is shown in enlarged scale in the upper right image section for each coating. While a rather confluent cell layer of VECs was only found on fibronectin-coated grafts, EPCs showed a rather confluent growth on all coatings and also uncoated grafts. Alb—albumin, Fib—fibronectin, Hep—heparin, Unc—uncoated.

3.3.2. Graft Colonization under Dynamic Conditions: Acridine Orange Staining

Under dynamic conditions, partially confluent areas of VECs were found for both the albumin and fibronectin groups, while the uncoated group showed a lower density of VECs (Figure 5 and Figure S3). The heparin group showed only scattered signals at much lower density. The most confluent coverage and highest density was clearly demonstrated by the fibronectin group.

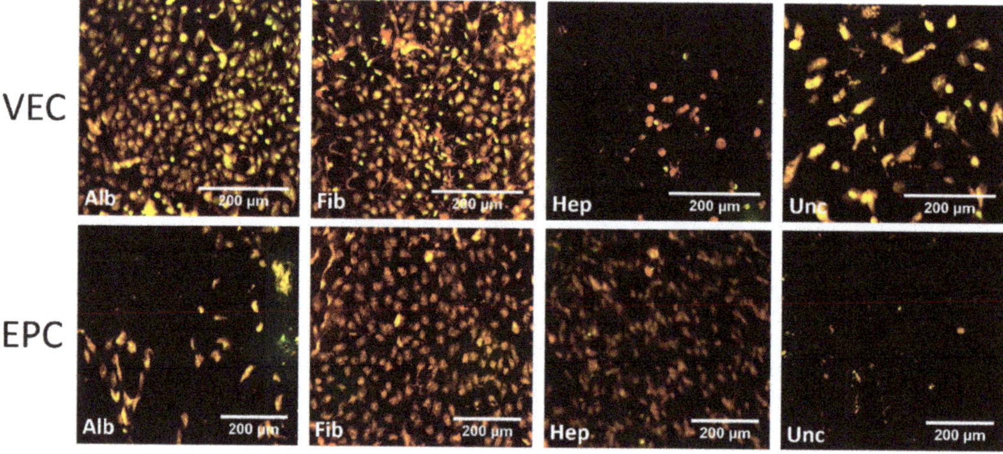

Figure 5. The image shows the luminal side of a sliced and spread out section of the tubular BNC graft seeded with vascular endothelial cells (VECs, upper row) and endothelial progenitor cells (EPCs, bottom row) after 96 h of dynamic cultivation, stained with acridine orange. All images reflect the exemplary regions that are labelled in Figures S3 and S4. Islands with high cell density of VECs were only found on albumin- and fibronectin-coated grafts, with lower density on uncoated grafts. The EPCs grew to highly dense populated areas only on fibronectin- and heparin-coated grafts, whereas the albumin and uncoated groups only showed isolated cells and cell groups. Alb—albumin, Fib—fibronectin, Hep—heparin, Unc—uncoated.

Colonization by EPCs resulted in an apparently similar coverage for the fibronectin group (Figure 5 and Figure S4). Here, the heparin group was also colonized, but at lower cell densities. The albumin and uncoated groups remained mostly unpopulated, and only scattered cells were found.

3.3.3. Cytoskeleton—Static Conditions

The F-actin staining in Figure 6 shows the cytoskeleton of the endothelial cells that were cultured on BNC patches for 96 h. For VECs, apparent stress fibers could be observed mainly in the albumin and fibronectin groups, while the heparin and uncoated groups showed rather diffuse and rare F-Actin signals, concentrated around the cell nucleus. Therefore, the albumin group showed denser and more oriented signals compared to the fibronectin group. In contrast to the VECs, apparent stress fibers could be detected in all groups colonized by EPCs. Here, F-Actin signals were generally at a higher intensity, showing denser stress fibers, which seemed to be almost equally distributed across the groups. However, the fibronectin group showed the highest density of apparent fibers, but also at the highest cell density. As apparent in the heparin group and less intensive in the fibronectin group, the fibers were more concentrated at the cell margin compared to the cytoplasmic area.

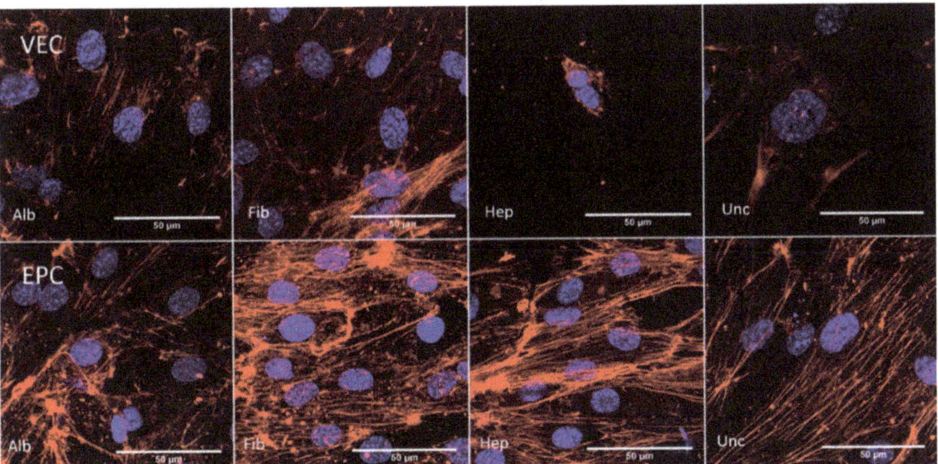

Figure 6. Confocal microscopy images of BNC grafts seeded with vascular endothelial cells (VEC) and endothelial progenitor cells (EPC) under static conditions. Cell nuclei were stained against DAPI (blue) and the F-actin of the cytoskeleton was stained with rhodamine phalloidin (red). All groups seeded with EPCs showed areas with dense colonization and a distinct cytoskeleton. For VECs, it was only in the heparin group that singular cells without pronounced a cytoskeleton were found, while VECs in the other groups showed areas of cells with a regular expansion of the F-actin filaments.

3.3.4. Cytoskeleton—Dynamic Conditions

The pattern of F-actin filaments under dynamic conditions is shown in Figure 7. Visible fibers were detected for all groups but heparin, where cells apparently were not vital. For the albumin and fibronectin groups, fibers were densely concentrated at the cell margin. While the uncoated group showed a less dense network of F-actin filaments, cells in the heparin group were only detected as scattered, singular and apparently not vital cells. This pattern of the cytoskeleton was also seen for EPCs in the uncoated and albumin groups. In contrast to this, the stress fibers of EPCs in the fibronectin group showed a strong signal, and singular fibers as well as fiber bundles at the cell margin were visible. The heparin group also showed a distinct F-actin signal, while the fibers were not visible as singular strands but rather as concentrated bundles at the cell margin.

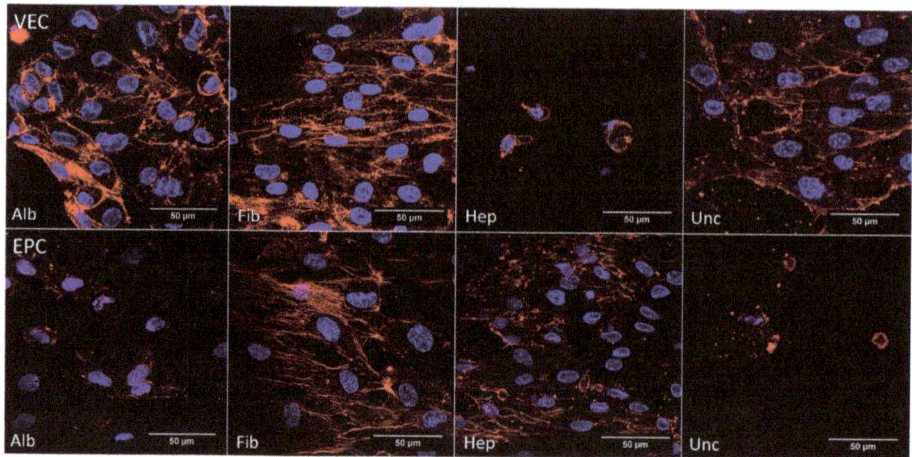

Figure 7. Confocal microscopy images of BNC grafts seeded with vascular endothelial cells (VEC) and endothelial progenitor cells (EPC) under dynamic conditions. Cell nuclei were stained against DAPI (blue) and the F-actin of the cytoskeleton was stained with rhodamine phalloidin (red). The fibronectin group showed a pronounced expansion of the cytoskeleton for both cell types. The VECs also showed a distinct expression of F-actin filaments on uncoated grafts. For the albumin group, the F-actin filaments were concentrated around the nuclei, indicating less spread of the cell bodies, and the heparin group showed no distinct signal of F-actin filaments. For EPCs, cells in the albumin and uncoated groups did not show clearly recognizable structures of F-actin filaments. EPCs on heparin-coated grafts showed a distinct F-actin signal.

3.3.5. CD31—Dynamic Conditions

The images obtained from parts of the BNC grafts under dynamic culture and stained against CD31 are shown in Figure 8. For VECs, a distinct signal of CD31 on the cell surface was only found for the fibronectin group, and only partially in the albumin group. In the heparin group, there were hardly any vital cells with recognizable structures and clear CD31 signals. For EPCs, a clear signal was found for the fibronectin and heparin groups, but not for the albumin and uncoated groups.

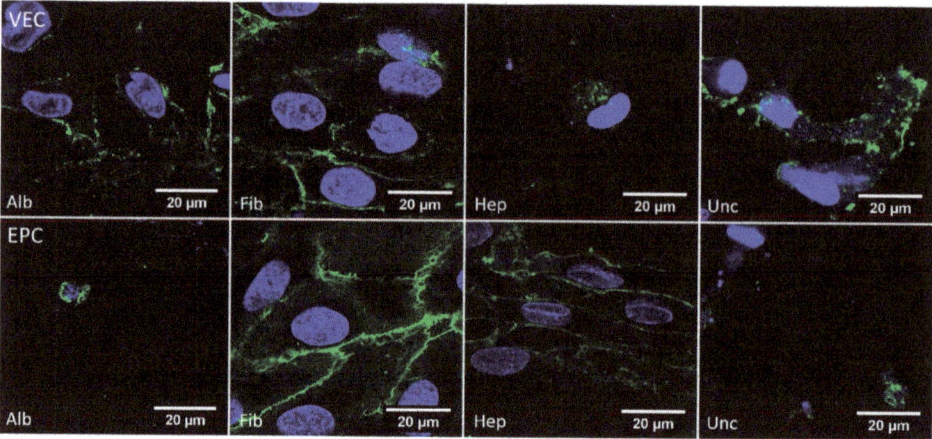

Figure 8. Confocal microscopy images of BNC grafts seeded with vascular endothelial cells cultivated under dynamic conditions and stained against DAPI (blue) and CD31 (green). A distinct signal of CD31 on VECs was only seen clearly in the fibronectin group, and for EPCs in the fibronectin and heparin groups.

3.4. AcLDL Assay

3.4.1. Static Conditions

The results of the AcLDL uptake assay are shown in Figure 9. In the albumin, heparin, and uncoated groups, no intracellular AcLDL particles were found, while cells in the fibronectin group partially showed uptake of the AcLDL particles to the cytoplasm. In contrast to this, the EPCs showed uptake of AcLDL particles for all groups.

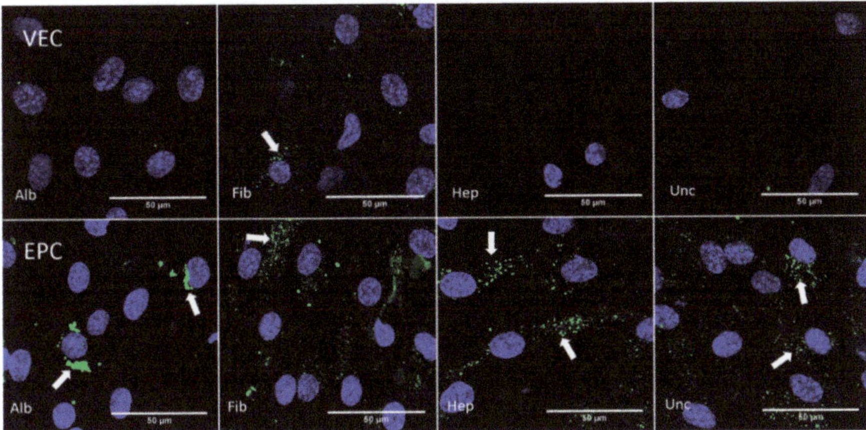

Figure 9. Confocal microscopy images from vascular endothelial cells (VEC) and endothelial progenitor cells (EPC) cultivated under static conditions on bacterial nanocellulose patches coated with albumin (Alb), fibronectin (Fib), heparin–chitosan (Hep) and on uncoated (Unc) patches. The cell nuclei were stained against DAPI (blue), and acetylated low density lipoprotein particles are shown in green (indicated by arrows). The VECs in the albumin and uncoated groups only showed nonspecific signals.

3.4.2. Dynamic Conditions

Under dynamic conditions, partial uptake of AcLDL particles into VECs was only seen in the fibronectin group, while the other groups did not show any clear signals (Figure 10). For EPCs, intracellular AcLDL particles were present mainly for the fibronectin and heparin groups, while no specific signal was found in the albumin and uncoated groups.

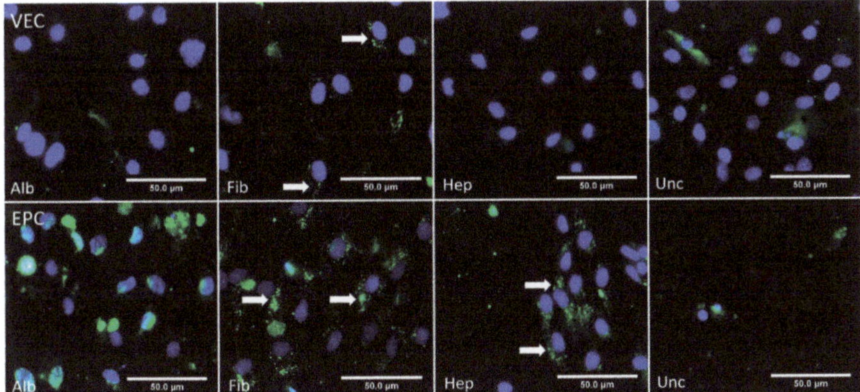

Figure 10. Fluorescence microscopy images from vascular endothelial cells (VEC) and endothelial progenitor cells (EPC) cultured under dynamic conditions on bacterial nanocellulose grafts coated with albumin (Alb), fibronectin (Fib), heparin–chitosan (Hep) and on uncoated (Unc) grafts. The cell nuclei were stained against DAPI (blue), and acetylated low density lipoprotein particles are shown in green (indicated by arrows).

4. Discussion

Endothelialization of small-diameter vascular grafts remains to be a crucial factor for both short- and long-term patencies, exhibited by several animal models [54–57] and also in human coronary artery bypass grafting [58,59]. This is most appealing for acellular off-the-shelf grafts, which exhibit on-site endothelialization after implantation. From the clinical perspective, the preservation of endothelial function in autologous vessels is considered to be indispensable for long-term patency [58,59]. Bacterial nanocellulose small-diameter grafts are thought to be a promising approach in the development of acellular grafts for in vivo population with ECs [14,16,17,34,60]. However, in contrast to the native extracellular matrix, BNC scaffolds fail to provide tissue-specific proteins for cell adhesion.

Accordingly, in this study, we used an uncoated BNC graft during the in vitro dynamic experimental setup, which proved to be challenging for patient-derived endothelial cells. The uncoated grafts showed poor EC growth kinetics in the WST-1 assay except for EPCs under static conditions, and only scattered cells were found, as evidenced by nucleic acids stains such as acridine orange. The morphology of EPCs was impaired as evidenced by F-actin staining, and apparently, endothelial functionality could not be preserved, as indicated by a deficiency in CD31 signals and active AcLDL uptake by both cell types. We have therefore provided an additional confirmation to show the poor performance of human microvascular ECs [61] and isolated human saphenous vein ECs [34] on uncoated grafts, which were in line with previous reports. In these studies, the authors showed that ECs seeded on unmodified BNC surfaces exhibit low adhesion and proliferation and do not spread but remain in a round shape, without the formation of a typical endothelial cell-like cytoskeleton.

Here, we contrasted the effects of coating BNC-SDVGs with different substances of previously reported relevance with regard to endothelialization in a setup using BNC-SDVGs of clinically relevant length and diameter. Thereby, we hoped to find an optimal candidate regarding adhesion and growth of clinically relevant endothelial cell types. We hypothesized that coating of small-diameter BNC grafts of clinically relevant length and diameter with one of these substances would increase the adhesion and vitality of clinical relevant EC types, such as human saphenous vein ECs (HSVECs) or EPCs. Special emphasis was given to the biological model of these two cell types, which are of particular interest because the in vivo endothelialization of vascular prosthesis depends mainly on two mechanisms: (I) the continuous ingrowth of ECs at the anastomic site, which are derived by proliferation of VECs but are often restricted to a length of 1–2 cm; (II) colonization of more distended regions by EPCs from peripheral blood [62].

Adhesion and anchorage of cells on an artificial surface is a complex process. Therefore, and though not analyzed in this study, it should also be considered that next to specific ligand binding, cell adhesion is also impacted by the biomaterial's hydrophilicity and surface charge [63].

Regarding the surface charge, for the in vivo situation, a positive surface charge is considered to indirectly enhance cell adhesion and proliferation by enabling adsorption of plasmatic proteins, such as fibronectin, which in turn can provide specific domains for anchorage [63–65]. However, in vitro, it is rather the density of charges that fosters unspecific affinities to the cell adhesion molecules [66].

Unmodified BNC has been shown to have a negative charge [65], and cell adhesion could be promoted by introduction of a positive ionic charge [65,67]. Data on the surface charge of fibronectin-, albumin- or heparin–chitosan-modified BNC surfaces is limited, but fibronectin has been shown to bear a mild negative potential under experimental conditions. In contrast, albumin [68,69] and heparin [70] are both considered to be highly negatively charged molecules, while chitosan is considered to be a positively charged molecule [71,72].

As a measure for hydrophilicity, the water contact angle became a common parameter to estimate the potential for a general adhesion affinity [63]. It has been shown that the coating of biomaterials with fibronectin [21] or its binding site RGD [33], heparin [73], and albumin [27] impacts the water contact angle, where heparin- and albumin-modified surfaces showed reduced water contact angle measurements, but heparin–chitosan-modified BNC was characterized as more hydrophilic compared to native BNC [22]. Thus, regarding the expected functionality for endothelialization, the role of fibronectin as a protein of the extracellular matrix that binds via its RGD binding domain to integrins of the EC membrane (e.g., $\alpha V\beta 3$ and $\alpha IIb\beta 3$), allowing adhesion of the cells and thereby increasing the cell viability, is well established and widely used [35,74,75].

In contrast, mechanisms for cell adhesion on albumin seems to be rather unspecific, as integrin binding domains have not been identified for albumin yet [75]. Here, increasing the hydrophilicity of albumin-coated surfaces could improve cell adhesion, mediated by the previously mentioned indirect mechanisms by the recruitment of plasmatic proteins that are currently under debate [27,63,76].

Regarding heparin–chitosan, ECs lack a direct receptor for heparin, but it is speculated that the glucosamine group of chitosan contains domains that interact with integrins and receptors of the cell membrane, thereby modulating cell spread and proliferation [32].

After all, coating procedures for all substances could be successfully transferred to our grafts by established methodology [21,22]. After seeding and biomimetic perfusion culture in our bioreactor, the fibronectin coating clearly outperformed both albumin and heparin, as well as the uncoated BNC. Both VECs and EPCs efficiently adhered to the graft and formed densely populated colonies with preserved endothelial functions, and in addition, also built stress fibers. In the WST-1 assay, the fibronectin-coated grafts achieved approximately one third of the metabolic activity of the control groups for both cell types after a culture time of 96 h, under both static and dynamic conditions, indicating that the cells adhered well and were not washed out under flow conditions for longer hours. This was further evidenced by a dense cell population with acridine orange staining, and further, we found typical expression of stress fibers and functional integrity as evidenced by positive CD31 signals and AcLDL uptake by the cells, respectively. Importantly, this accounted for both cell types and both static and dynamic conditions, making the results most consistent compared to the other coatings, where cell adhesion and vitality differed between cell types and static or dynamic conditions.

Generally, this is in line with the established role of fibronectin as a protein of the extracellular matrix that binds via its RGD binding domain to integrins of the EC membrane, allowing adhesion of the cells and thereby increasing the cell viability [74,75]. Besides ECs, fibronectin also increases the proliferation of other SDVG-relevant cell types, such as vascular smooth muscle cells, leading to intimal hyperplasia and narrowing of the internal vessel lumen [77,78]. Previous studies with in vivo experiments have not shown excessive intimal hyperplasia in fibronectin coated SDVGs [79–81], but some studies have already shown that fibronectin coating increases EC adhesion to acellular vascular grafts [81–85] and also to BNC [21].

Kuzmenko et al. found that coating of CDAP-activated BNC with fibronectin leads to improved endothelialization with HUVECs under static conditions [21]. Recently, Osorio et al. used decellularized BNC constructs that were previously seeded with fibroblasts and showed that the BNC surface was modified with a fibronectin-containing extracellular matrix, resulting in higher cell populations in the following re-seeding experiments [86]. Consistent with the VECs that we used, Bodin et al. used HSVECs to study the repopulation of BNC modified with the fibronectin cell-binding domain RGD [33]. They found increased endothelialization for the modified BNC constructs in comparison to uncoated constructs and concluded that the RGD modification was the key factor for the improved cell adherence. We observed similar behaviors of EPCs on our fibronectin-coated grafts, which is also in line with the literature, as it is known that fibronectin coating of vascular grafts induces the homing of CD34+ endothelial progenitor cells [87] and promotes VEGF-induced $CD34^+$ cell differentiation into ECs [88]. Consistent with that, our EPCs were characterized as $CD34^+$ by FACS.

Importantly, it should also be considered that fibronectin coating of vascular prostheses could have unfavorable effects with regard to thrombogenicity. Fibronectin-modified surfaces are able to capture platelets via β1 and αIIbβ3 integrins, probably resulting in increased thrombogenicity [89–92], which might be counterweighted by platelet-derived EC growth factors, which in turn stimulate the growth of endothelial cells [77,93], which has already been shown by Stronck et al. with regard to endothelialization of SDVGs [94]. Also, platelet adhesion after vascular interventions is controlled by the standard administration of antiplatelet drugs [95,96], which might further lower the prothrombogenic effect of fibronectin coating.

Nevertheless, heparin and heparin–chitosan coating have also generally been shown to enhance the proliferation of ECs in vitro [22,97]. For example, Li et al. found a significant proliferation of porcine iliac ECs on heparin–chitosan-modified BNC grafts under static culture conditions [22]. Contradictorily, we found no reasonable VEC growth as the metabolic activity ranged even below uncoated BNC grafts for both static and dynamic conditions. However, the reason behind the lack of EC growth might be due to the cell type that were used in this study. Heparin has previously been shown to inhibit the proliferation of endothelial cells, particularly HSVECs [98–100]. Furthermore, heparin–chitosan-coated surfaces did not support the proliferation of other clinically relevant primary ECs, such as human coronary ECs [32].

Though albumin coating has been studied intensively as a candidate for the functionalization of the surfaces of medical devices in the last 30 years, data on albumin coating of vascular prostheses is limited. Importantly, it has been shown that human serum albumin inhibits apoptosis in ECs [101] and can also act as an anti-inflammatory agent by inhibiting TNFα-induced VCAM-1 expression and monocyte adhesion in human aortic ECs [102]. Furthermore, it is known that serum albumin immobilization on blood-contacting materials reduces thrombogenicity because adhesion of platelets and leukocytes is reduced [103,104]. These attributes of albumin still mark albumin as an interesting candidate for the coating of vascular prostheses.

However, in this study, VEC proliferation activity on albumin-coated grafts was only 10% of the controls under static conditions after 96 h, while up to 33% was achieved under dynamic conditions (Table S1) as measured by the WST-1 assay. Interestingly, this was flipped for EPCs, where no significant growth was seen under dynamic conditions, indicating that the cells did not adhere properly or might have been washed away during perfusion. In addition, the few adhered cells were functionally impaired, as evidenced by their inability to take up the AcLDL particles. Interestingly, in the literature, the impact of an albumin coating on endothelial cell growth is assessed ambiguously. In vivo, a bovine serum albumin aptamer coating of titanium discs, implanted in the iliac artery of dogs, enhanced proliferation of canine EPCs but not vascular ECs [27], while albumin coating of vascular Dacron prostheses did not enhance the in vivo endothelialization in another canine model [105]. In contrast to this, Krajewski et al. found that albumin-coated stents showed improved endothelialization with HUVECs [28], while others found reduced endothelialization for albumin-coated polyester grafts [29] or albumin-coated polyethylene (PET) vascular prostheses [27], when both seeded with HUVECs.

In direct comparison, EPCs exhibited improved performance for all tested parameters, even in the heparin-coated grafts. They showed reasonable growth under static conditions and were apparently not washed away under perfusion, as the metabolic activity reached approximately 30% of the controls after 96 h of dynamic cultivation. We found dense colonized areas, which is a reasonable signal of stress fibers and preserved functional integrity with respect to positive AcLDL signals.

The basic mechanism behind the different growth pattern for VECs in contrast to EPCs that were observed for the albumin- and heparin-coated grafts remains obscure. Perhaps these differences could be explained by the differences in cell proliferation potential and gene expression profile of ECs obtained from different localizations. It has been shown that EPCs provide more robust adhesion characteristics due to highly expressed genes associated with proliferation, adhesion, and motility (VEGFR1, JMJD6, and TGFB3) and indeed, they multiply rapidly in comparison to saphenous vein ECs and are more resistant to apoptosis [42,106–108]. So, heparin–chitosan coating still might enable colonization by EPC in vivo.

Cumulatively, the results obtained from the three different types of coated grafts reinforces the fact that except for the albumin-coated graft, where the results are still indefinite, the coating of BNC grafts with either fibronectin or heparin–chitosan remains most promising for the future development of small-diameter vascular grafts.

Taken together, this study indicates that endothelialization by the capturing of endothelial progenitors seems possible for heparin–chitosan-coated BNC grafts. Thereby, heparin–chitosan-coated grafts would be a very interesting option for small-diameter vascular grafts, even if the continuous ingrowth of ECs from the anastomic region could be impaired. It is known that intimal hyperplasia caused by smooth muscle cells at the anastomosis is one of the causes for graft failure [109,110], and interestingly, Chupa et al. found impaired proliferation and cell spread of smooth muscle cells on heparin–chitosan-modified surfaces [32], which is also confirmed by the findings of Clowes et al. [111]. Thus, a heparin–chitosan coating could probably limit the intimal hyperplasia at the anastomosis to the native vessel, while promoting endothelialization with EPCs originating from the donor blood, thereby overcoming the disadvantage of reduced attachment of HSVECs.

In summary, our data indicate that in our setting, only the groups containing integrin-binding domains—presented here by fibronectin, and potentially by chitosan—enabled colonization with preserved endothelial functionality. The albumin group revealed adhesion, but no clear signals for AcLDL uptake, and CD31 expression was low on VEC and absent on EPC.

5. Conclusions

In this study, we could confirm that for both VECs and EPCs, bacterial nanocellulose per se does not seem to be an attractive material for the formation of a tissue-typical endothelium. Coating with fibronectin clearly outperformed albumin and heparin–chitosan coating, regarding cell growth and preservation of endothelial functions for both cell lines. But heparin–chitosan-coated BNC grafts still allowed the attachment and spread of EPCs. Hence, for off-the-shelf small-diameter vascular grafts made of BNC, coating with either fibronectin or heparin–chitosan might promote the important endothelialization by host-cells. Further optimization is warranted and should be addressed in detail with in vivo settings.

Supplementary Materials: The following are available online at https://www.mdpi.com/article/10.3390/nano11081952/s1: Figure S1: Exemplary characterization of vascular endothelial cells (VEC) and endothelial progenitor cells (EPC) by flow cytometry. Figure S2: Exemplary characterization of vascular endothelial cells (VEC) and endothelial progenitor cells (EPC). Figure S3: Stitched microscopy images of the luminal side of a sliced and spread out section of the tubular BNC graft seeded with vascular endothelial cells after 96 h of dynamic cultivation, stained with acridine orange. Figure S4: Stitched microscopy images of the luminal side of a sliced and spread out section of the tubular BNC graft seeded with endothelial progenitor cells after 96 h of dynamic cultivation, stained with acridine orange. Table S1: Results of WST-1 proliferation assay performed on the cell seeded BNC grafts under both static and dynamic conditions for vascular endothelial cells and endothelial progenitor cells. Table S2: Protein coating stability. During the coating procedure of albumin and fibronectin patches, samples from the respective coating solution before coating and after the incubation time of 12 h were obtained, as well as from the PBS in which the samples were stored.

Author Contributions: Conceptualization, M.W., J.H., H.W., J.W., and M.S.; methodology, M.W., J.H., J.R., H.W., S.S., B.G., and P.V.; software, M.W., J.R. and S.V.; formal analysis, M.W., J.R., J.H., and G.A.; investigation, M.W., J.R.; writing—original draft preparation, M.W., J.H. and J.R., writing—review and editing, M.W., G.A., S.S., P.V., and S.V.; supervision, H.W. and S.S.; funding acquisition, M.W., J.W., H.W., and M.S. All authors have read and agreed to the published version of the manuscript.

Funding: This project was funded by the German Research Foundation (grant number WA 4489/1-1).

Institutional Review Board Statement: The study was conducted according to the guidelines of the Declaration of Helsinki, and approved by the Institutional Ethics Committee of the University Hospital of Magdeburg (application number 81/18, approved 9 July 2018).

Informed Consent Statement: Informed consent was obtained from all subjects involved in the study.

Data Availability Statement: The data presented in this study are available in this manuscript and the supplementary data.

Acknowledgments: Special thanks go to Torsten Doenst (University Hospital Jena, Department of Cardiothoracic Surgery) for his support during the planning of the project within the framework of the "Nachwuchsakademie Herzchirurgie 2018", to Elena Denks for her technical assistance, to rer. nat. Roland Hartig for his support in confocal microscopy, and to Peter Knüppel for his assistance with the development of the customized rotation unit for cell seeding.

Conflicts of Interest: The authors declare no conflict of interest. The funders had no role in the design of the study; in the collection, analyses, or interpretation of data; in the writing of the manuscript; or in the decision to publish the results.

References

1. Diodato, M.; Chedrawy, E.G. Coronary artery bypass graft surgery: The past, present, and future of myocardial revascularisation. *Surg. Res. Pract.* **2014**, *2014*, 726158. [CrossRef]
2. Beckmann, A.; Funkat, A.K.; Lewandowski, J.; Frie, M.; Ernst, M.; Hekmat, K.; Schiller, W.; Gummert, J.F.; Cremer, J.T. Cardiac Surgery in Germany during 2014: A Report on Behalf of the German Society for Thoracic and Cardiovascular Surgery. *Thorac. Cardiovasc. Surg.* **2015**, *63*, 258–269. [CrossRef] [PubMed]
3. Martinez-Gonzalez, B.; Reyes-Hernandez, C.G.; Quiroga-Garza, A.; Rodriguez-Rodriguez, V.E.; Esparza-Hernandez, C.N.; Elizondo-Omana, R.E.; Guzman-Lopez, S. Conduits Used in Coronary Artery Bypass Grafting: A Review of Morphological Studies. *Ann. Thorac. Cardiovasc. Surg.* **2017**, *23*, 55–65. [CrossRef] [PubMed]
4. He, G.W. Arterial grafts for coronary artery bypass grafting: Biological characteristics, functional classification, and clinical choice. *Ann. Thorac. Surg.* **1999**, *67*, 277–284. [CrossRef]
5. Gaudino, M.; Glieca, F.; Luciani, N.; Pragliola, C.; Tsiopoulos, V.; Bruno, P.; Farina, P.; Bonalumi, G.; Pavone, N.; Nesta, M.; et al. Systematic bilateral internal mammary artery grafting: Lessons learned from the CATHolic University EXtensive BIMA Grafting Study. *Eur. J. Cardiothorac. Surg.* **2018**, *54*, 702–707. [CrossRef]
6. Wang, X.; Lin, P.; Yao, Q.; Chen, C. Development of small-diameter vascular grafts. *World J. Surg.* **2007**, *31*, 682–689. [CrossRef]
7. Piccone, V. Alternative techniques in coronary artery reconstruction. In *Modern Vascular Grafts*; PN, S., Ed.; McGraw-Hill: New York, NY, USA, 1987; pp. 253–260.
8. Gaudino, M.; Antoniades, C.; Benedetto, U.; Deb, S.; Di Franco, A.; Di Giammarco, G.; Fremes, S.; Glineur, D.; Grau, J.; He, G.W.; et al. Mechanisms, Consequences, and Prevention of Coronary Graft Failure. *Circulation* **2017**, *136*, 1749–1764. [CrossRef]
9. Pashneh-Tala, S.; MacNeil, S.; Claeyssens, F. The Tissue-Engineered Vascular Graft-Past, Present, and Future. *Tissue Eng. Part B Rev.* **2015**, *22*, 68–100. [CrossRef] [PubMed]
10. Goh, E.T.; Wong, E.; Farhatnia, Y.; Tan, A.; Seifalian, A.M. Accelerating in situ endothelialisation of cardiovascular bypass grafts. *Int. J. Mol. Sci.* **2014**, *16*, 597–627. [CrossRef] [PubMed]
11. Patel, S.D.; Waltham, M.; Wadoodi, A.; Burnand, K.G.; Smith, A. The role of endothelial cells and their progenitors in intimal hyperplasia. *Ther. Adv. Cardiovasc. Dis.* **2010**, *4*, 129–141. [CrossRef] [PubMed]
12. Neumann, F.J.; Sousa-Uva, M.; Ahlsson, A.; Alfonso, F.; Banning, A.P.; Benedetto, U.; Byrne, R.A.; Collet, J.P.; Falk, V.; Head, S.J.; et al. 2018 ESC/EACTS Guidelines on myocardial revascularization. *EuroIntervention* **2019**, *14*, 1435–1534. [CrossRef] [PubMed]
13. Radke, D.; Jia, W.; Sharma, D.; Fena, K.; Wang, G.; Goldman, J.; Zhao, F. Tissue Engineering at the Blood-Contacting Surface: A Review of Challenges and Strategies in Vascular Graft Development. *Adv. Healthc. Mater.* **2018**, *7*, e1701461. [CrossRef] [PubMed]
14. Scherner, M.; Reutter, S.; Klemm, D.; Sterner-Kock, A.; Guschlbauer, M.; Richter, T.; Langebartels, G.; Madershahian, N.; Wahlers, T.; Wippermann, J. In vivo application of tissue-engineered blood vessels of bacterial cellulose as small arterial substitutes: Proof of concept? *J. Surg. Res.* **2014**, *189*, 340–347. [CrossRef]
15. Wippermann, J.; Schumann, D.; Klemm, D.; Kosmehl, H.; Salehi-Gelani, S.; Wahlers, T. Preliminary results of small arterial substitute performed with a new cylindrical biomaterial composed of bacterial cellulose. *Eur. J. Vasc. Endovasc. Surg.* **2009**, *37*, 592–596. [CrossRef] [PubMed]
16. Weber, C.; Reinhardt, S.; Eghbalzadeh, K.; Wacker, M.; Guschlbauer, M.; Maul, A.; Sterner-Kock, A.; Wahlers, T.; Wippermann, J.; Scherner, M. Patency and in vivo compatibility of bacterial nanocellulose grafts as small-diameter vascular substitute. *J. Vasc. Surg.* **2018**, *68*, 177S–187S. [CrossRef] [PubMed]

17. Wacker, M.; Kiesswetter, V.; Slottosch, I.; Awad, G.; Paunel-Gorgulu, A.; Varghese, S.; Klopfleisch, M.; Kupitz, D.; Klemm, D.; Nietzsche, S.; et al. In vitro hemo- and cytocompatibility of bacterial nanocelluose small diameter vascular grafts: Impact of fabrication and surface characteristics. *PLoS ONE* **2020**, *15*, e0235168. [CrossRef]
18. Zhuang, Y.; Zhang, C.; Cheng, M.; Huang, J.; Liu, Q.; Yuan, G.; Lin, K.; Yu, H. Challenges and strategies for in situ endothelialization and long-term lumen patency of vascular grafts. *Bioact. Mater.* **2021**, *6*, 1791–1809. [CrossRef]
19. Avci-Adali, M.; Perle, N.; Ziemer, G.; Wendel, H.P. Current concepts and new developments for autologous in vivo endothelialisation of biomaterials for intravascular applications. *Eur. Cell Mater.* **2011**, *21*, 157–176. [CrossRef] [PubMed]
20. Kumar, V.B.S.; Viji, R.I.; Kiran, M.S.; Sudhakaran, P.R. *Angiogenic Response of Endothelial Cells to Fibronectin*; Springer: New York, NY, USA, 2012; pp. 131–151.
21. Kuzmenko, V.; Samfors, S.; Hagg, D.; Gatenholm, P. Universal method for protein bioconjugation with nanocellulose scaffolds for increased cell adhesion. *Mater. Sci. Eng. C Mater. Biol. Appl.* **2013**, *33*, 4599–4607. [CrossRef]
22. Li, X.; Tang, J.; Bao, L.; Chen, L.; Hong, F.F. Performance improvements of the BNC tubes from unique double-silicone-tube bioreactors by introducing chitosan and heparin for application as small-diameter artificial blood vessels. *Carbohydr. Polym.* **2017**, *178*, 394–405. [CrossRef]
23. Bao, L.; Hong, F.F.; Li, G.; Hu, G.; Chen, L. Improved Performance of Bacterial Nanocellulose Conduits by the Introduction of Silk Fibroin Nanoparticles and Heparin for Small-Caliber Vascular Graft Applications. *Biomacromolecules* **2021**, *22*, 353–364. [CrossRef]
24. Follmann, H.D.; Naves, A.F.; Martins, A.F.; Felix, O.; Decher, G.; Muniz, E.C.; Silva, R. Advanced fibroblast proliferation inhibition for biocompatible coating by electrostatic layer-by-layer assemblies of heparin and chitosan derivatives. *J. Colloid Interface Sci.* **2016**, *474*, 9–17. [CrossRef]
25. Wang, Y.; He, C.; Feng, Y.; Yang, Y.; Wei, Z.; Zhao, W.; Zhao, C. A chitosan modified asymmetric small-diameter vascular graft with anti-thrombotic and anti-bacterial functions for vascular tissue engineering. *J. Mater. Chem. B* **2020**, *8*, 568–577. [CrossRef]
26. Aussel, A.; Thebaud, N.B.; Berard, X.; Brizzi, V.; Delmond, S.; Bareille, R.; Siadous, R.; James, C.; Ripoche, J.; Durand, M.; et al. Chitosan-based hydrogels for developing a small-diameter vascular graft: In vitro and in vivo evaluation. *Biomed. Mater.* **2017**, *12*, 065003. [CrossRef]
27. Chen, Z.; Li, Q.; Chen, J.; Luo, R.; Maitz, M.F.; Huang, N. Immobilization of serum albumin and peptide aptamer for EPC on polydopamine coated titanium surface for enhanced in-situ self-endothelialization. *Mater. Sci. Eng. C Mater. Biol. Appl.* **2016**, *60*, 219–229. [CrossRef] [PubMed]
28. Krajewski, S.; Neumann, B.; Kurz, J.; Perle, N.; Avci-Adali, M.; Cattaneo, G.; Wendel, H.P. Preclinical evaluation of the thrombogenicity and endothelialization of bare metal and surface-coated neurovascular stents. *AJNR Am. J. Neuroradiol.* **2015**, *36*, 133–139. [CrossRef] [PubMed]
29. Marois, Y.; Chakfe, N.; Guidoin, R.; Duhamel, R.C.; Roy, R.; Marois, M.; King, M.W.; Douville, Y. An albumin-coated polyester arterial graft: In vivo assessment of biocompatibility and healing characteristics. *Biomaterials* **1996**, *17*, 3–14. [CrossRef]
30. Tan, P.H.; Chan, C.; Xue, S.A.; Dong, R.; Ananthesayanan, B.; Manunta, M.; Kerouedan, C.; Cheshire, N.J.; Wolfe, J.H.; Haskard, D.O.; et al. Phenotypic and functional differences between human saphenous vein (HSVEC) and umbilical vein (HUVEC) endothelial cells. *Atherosclerosis* **2004**, *173*, 171–183. [CrossRef]
31. Haug, V.; Torio-Padron, N.; Stark, G.B.; Finkenzeller, G.; Strassburg, S. Comparison between endothelial progenitor cells and human umbilical vein endothelial cells on neovascularization in an adipogenesis mouse model. *Microvasc. Res.* **2015**, *97*, 159–166. [CrossRef] [PubMed]
32. Chupa, J.M.; Foster, A.M.; Sumner, S.R.; Madihally, S.V.; Matthew, H.W. Vascular cell responses to polysaccharide materials: In vitro and in vivo evaluations. *Biomaterials* **2000**, *21*, 2315–2322. [CrossRef]
33. Bodin, A.; Ahrenstedt, L.; Fink, H.; Brumer, H.; Risberg, B.; Gatenholm, P. Modification of nanocellulose with a xyloglucan-RGD conjugate enhances adhesion and proliferation of endothelial cells: Implications for tissue engineering. *Biomacromolecules* **2007**, *8*, 3697–3704. [CrossRef] [PubMed]
34. Fink, H.; Ahrenstedt, L.; Bodin, A.; Brumer, H.; Gatenholm, P.; Krettek, A.; Risberg, B. Bacterial cellulose modified with xyloglucan bearing the adhesion peptide RGD promotes endothelial cell adhesion and metabolism—A promising modification for vascular grafts. *J. Tissue. Eng. Regen. Med.* **2011**, *5*, 454–463. [CrossRef]
35. Jalali, S.; del Pozo, M.A.; Chen, K.; Miao, H.; Li, Y.; Schwartz, M.A.; Shyy, J.Y.; Chien, S. Integrin-mediated mechanotransduction requires its dynamic interaction with specific extracellular matrix (ECM) ligands. *Proc. Natl. Acad. Sci. USA* **2001**, *98*, 1042–1046. [CrossRef]
36. Klemm, D.; Marsch, S.; Schumann, D.; Udhardt, U. Method and Device for Producing Shaped Microbial Cellulose for Use as Biomaterial, Especially for Microsurgery. U.S. Patent US 2003/0013163 A1, 16 January 2003.
37. Klemm, D.; Kopsch, V.; Koth, D.; Kramer, F.; Moritz, S.; Richter, T.; Ulrike, U.; Fried, W. *Device, Used To Prepare Hollow Bodies Made Of Microbial Cellulose, Includes Template, A First Reservoir, Which Is Filled With A Mixture Comprising A Liquid Culture Medium And A Polymer-Forming Microorganism, A Wetting Device, And A Housing*; Deutsches Patent- und Markenamt: Jena, Germany, 2013.
38. Klemm, D.; Petzold-Welcke, K.; Kramer, F.; Richter, T.; Raddatz, V.; Fried, W.; Nietzsche, S.; Bellmann, T.; Fischer, D. Biotech nanocellulose: A review on progress in product design and today′s state of technical and medical applications. *Carbohydr. Polym.* **2021**, *254*, 117313. [CrossRef]
39. Hercules, D.M. Electron Spectroscopy: Applications for Chemical Analysis. *J. Chem. Educ.* **2004**, *81*, 1751. [CrossRef]

40. Hinrichs, W.L.J.; ten Hoopen, H.W.M.; Wissink, M.J.B.; Engbers, G.H.M.; Feijen, J. Design of a new type of coating for the controlled release of heparin. *J. Control. Release* **1997**, *45*, 163–176. [CrossRef]
41. Watkins, M.T.; Sharefkin, J.B.; Zajtchuk, R.; Maciag, T.M.; D'Amore, P.A.; Ryan, U.S.; Van Wart, H.; Rich, N.M. Adult human saphenous vein endothelial cells: Assessment of their reproductive capacity for use in endothelial seeding of vascular prostheses. *J. Surg. Res.* **1984**, *36*, 588–596. [CrossRef]
42. Ormiston, M.L.; Toshner, M.R.; Kiskin, F.N.; Huang, C.J.; Groves, E.; Morrell, N.W.; Rana, A.A. Generation and Culture of Blood Outgrowth Endothelial Cells from Human Peripheral Blood. *J. Vis. Exp.* **2015**, *106*, e53384. [CrossRef] [PubMed]
43. Niwa, K.; Kado, T.; Sakai, J.; Karino, T. The effects of a shear flow on the uptake of LDL and acetylated LDL by an EC monoculture and an EC-SMC coculture. *Ann. Biomed. Eng.* **2004**, *32*, 537–543. [CrossRef] [PubMed]
44. Shnyra, A.; Lindberg, A.A. Scavenger receptor pathway for lipopolysaccharide binding to Kupffer and endothelial liver cells in vitro. *Infect. Immun.* **1995**, *63*, 865–873. [CrossRef] [PubMed]
45. Voyta, J.C.; Via, D.P.; Butterfield, C.E.; Zetter, B.R. Identification and isolation of endothelial cells based on their increased uptake of acetylated-low density lipoprotein. *J. Cell Biol.* **1984**, *99*, 2034–2040. [CrossRef]
46. Melchiorri, A.J.; Bracaglia, L.G.; Kimerer, L.K.; Hibino, N.; Fisher, J.P. In Vitro Endothelialization of Biodegradable Vascular Grafts Via Endothelial Progenitor Cell Seeding and Maturation in a Tubular Perfusion System Bioreactor. *Tissue. Eng. Part. C Methods* **2016**, *22*, 663–670. [CrossRef] [PubMed]
47. Schuerlein, S.; Schwarz, T.; Krziminski, S.; Gatzner, S.; Hoppensack, A.; Schwedhelm, I.; Schweinlin, M.; Walles, H.; Hansmann, J. A versatile modular bioreactor platform for Tissue Engineering. *Biotechnol. J.* **2017**, *12*, 1600326. [CrossRef] [PubMed]
48. Hulsmann, J.; Aubin, H.; Kranz, A.; Godehardt, E.; Munakata, H.; Kamiya, H.; Barth, M.; Lichtenberg, A.; Akhyari, P. A novel customizable modular bioreactor system for whole-heart cultivation under controlled 3D biomechanical stimulation. *J. Artif. Organs* **2013**, *16*, 294–304. [CrossRef] [PubMed]
49. Zhang, K.; Li, J.A.; Deng, K.; Liu, T.; Chen, J.Y.; Huang, N. The endothelialization and hemocompatibility of the functional multilayer on titanium surface constructed with type IV collagen and heparin. *Colloids Surf. B Biointerfaces* **2013**, *108*, 295–304. [CrossRef]
50. Avari, H.; Rogers, K.A.; Savory, E. Quantification of Morphological Modulation, F-Actin Remodeling and PECAM-1 (CD-31) Re-distribution in Endothelial Cells in Response to Fluid-Induced Shear Stress under Various Flow Conditions. *J. Biomech. Eng.* **2019**, *141*, 041004-1–041004-12. [CrossRef] [PubMed]
51. Dick, M.; Jonak, P.; Leask, R.L. Statin therapy influences endothelial cell morphology and F-actin cytoskeleton structure when exposed to static and laminar shear stress conditions. *Life Sci.* **2013**, *92*, 859–865. [CrossRef]
52. Schindelin, J.; Arganda-Carreras, I.; Frise, E.; Kaynig, V.; Longair, M.; Pietzsch, T.; Preibisch, S.; Rueden, C.; Saalfeld, S.; Schmid, B.; et al. Fiji: An open-source platform for biological-image analysis. *Nat. Methods* **2012**, *9*, 676–682. [CrossRef]
53. Preibisch, S.; Saalfeld, S.; Tomancak, P. Globally optimal stitching of tiled 3D microscopic image acquisitions. *Bioinformatics* **2009**, *25*, 1463–1465. [CrossRef]
54. Hytonen, J.P.; Leppanen, O.; Taavitsainen, J.; Korpisalo, P.; Laidinen, S.; Alitalo, K.; Wadstrom, J.; Rissanen, T.T.; Yla-Herttuala, S. Improved endothelialization of small-diameter ePTFE vascular grafts through growth factor therapy. *Vasc. Biol.* **2019**, *1*, 1–9. [CrossRef]
55. Budd, J.S.; Allen, K.E.; Hartley, G.; Bell, P.R. The effect of preformed confluent endothelial cell monolayers on the patency and thrombogenicity of small calibre vascular grafts. *Eur. J. Vasc. Surg.* **1991**, *5*, 397–405. [CrossRef]
56. James, N.L.; Schindhelm, K.; Slowiaczek, P.; Milthorpe, B.; Graham, A.R.; Munro, V.F.; Johnson, G.; Steele, J.G. In vivo patency of endothelial cell-lined expanded polytetrafluoroethylene prostheses in an ovine model. *Artif. Organs* **1992**, *16*, 346–353. [CrossRef] [PubMed]
57. Hao, D.; Fan, Y.; Xiao, W.; Liu, R.; Pivetti, C.; Walimbe, T.; Guo, F.; Zhang, X.; Farmer, D.L.; Wang, F.; et al. Rapid endothelialization of small diameter vascular grafts by a bioactive integrin-binding ligand specifically targeting endothelial progenitor cells and endothelial cells. *Acta. Biomater.* **2020**, *108*, 178–193. [CrossRef]
58. Samano, N.; Geijer, H.; Liden, M.; Fremes, S.; Bodin, L.; Souza, D. The no-touch saphenous vein for coronary artery bypass grafting maintains a patency, after 16 years, comparable to the left internal thoracic artery: A randomized trial. *J. Thorac. Cardiovasc. Surg.* **2015**, *150*, 880–888. [CrossRef]
59. Dashwood, M.R.; Savage, K.; Tsui, J.C.; Dooley, A.; Shaw, S.G.; Fernandez Alfonso, M.S.; Bodin, L.; Souza, D.S. Retaining perivascular tissue of human saphenous vein grafts protects against surgical and distension-induced damage and preserves endothelial nitric oxide synthase and nitric oxide synthase activity. *J. Thorac. Cardiovasc. Surg.* **2009**, *138*, 334–340. [CrossRef]
60. Zang, S.; Zhang, R.; Chen, H.; Lu, Y.; Zhou, J.; Chang, X.; Qiu, G.; Wu, Z.; Yang, G. Investigation on artificial blood vessels prepared from bacterial cellulose. *Mater Sci. Eng. C Mater Biol. Appl.* **2015**, *46*, 111–117. [CrossRef] [PubMed]
61. Andrade, F.K.; Costa, R.; Domingues, L.; Soares, R.; Gama, M. Improving bacterial cellulose for blood vessel replacement: Functionalization with a chimeric protein containing a cellulose-binding module and an adhesion peptide. *Acta. Biomater.* **2010**, *6*, 4034–4041. [CrossRef] [PubMed]
62. Zilla, P.; Bezuidenhout, D.; Human, P. Prosthetic vascular grafts: Wrong models, wrong questions and no healing. *Biomaterials* **2007**, *28*, 5009–5027. [CrossRef]
63. Bacakova, L.; Filova, E.; Parizek, M.; Ruml, T.; Svorcik, V. Modulation of cell adhesion, proliferation and differentiation on materials designed for body implants. *Biotechnol. Adv.* **2011**, *29*, 739–767. [CrossRef]

64. Kim, J.; Kim, D.H.; Lim, K.T.; Seonwoo, H.; Park, S.H.; Kim, Y.R.; Kim, Y.; Choung, Y.H.; Choung, P.H.; Chung, J.H. Charged nanomatrices as efficient platforms for modulating cell adhesion and shape. *Tissue Eng. Part C Methods* **2012**, *18*, 913–923. [CrossRef]
65. Courtenay, J.C.; Johns, M.A.; Galembeck, F.; Deneke, C.; Lanzoni, E.M.; Costa, C.A.; Scott, J.L.; Sharma, R.I. Surface modified cellulose scaffolds for tissue engineering. *Cellulose* **2017**, *24*, 253–267. [CrossRef]
66. Lindl, T.; Gstraunthaler, G. *Zell- und Gewebekultur*, 6th ed.; Spektrum Akademischer Verlag: Berlin/Heidelberg, Germany, 2008.
67. Watanabe, K.; Eto, Y.; Takano, S.; Nakamori, S.; Shibai, H.; Yamanaka, S. A new bacterial cellulose substrate for mammalian cell culture. A new bacterial cellulose substrate. *Cytotechnology* **1993**, *13*, 107–114. [CrossRef] [PubMed]
68. Wiig, H.; Kolmannskog, O.; Tenstad, O.; Bert, J.L. Effect of charge on interstitial distribution of albumin in rat dermis in vitro. *J. Physiol.* **2003**, *550*, 505–514. [CrossRef] [PubMed]
69. Baler, K.; Martin, O.A.; Carignano, M.A.; Ameer, G.A.; Vila, J.A.; Szleifer, I. Electrostatic unfolding and interactions of albumin driven by pH changes: A molecular dynamics study. *J. Phys. Chem. B* **2014**, *118*, 921–930. [CrossRef]
70. Chokhawala, H.A.; Yu, H.; Chen, X. Enzymatic Approaches to O-Glycoside Introduction: Glycosyltransferases☆. In *Reference Module in Chemistry, Molecular Sciences and Chemical Engineering*; Elsevier: Amsterdam, The Netherlands, 2013.
71. Moraru, C.; Mincea, M.; Menghiu, G.; Ostafe, V. Understanding the Factors Influencing Chitosan-Based Nanoparticles-Protein Corona Interaction and Drug Delivery Applications. *Molecules* **2020**, *25*, 4758. [CrossRef] [PubMed]
72. Sami El-banna, F.; Mahfouz, M.E.; Leporatti, S.; El-Kemary, M.; Hanafy, N.A.N. Chitosan as a Natural Copolymer with Unique Properties for the Development of Hydrogels. *Appl. Sci.* **2019**, *9*, 2193. [CrossRef]
73. He, S.; Chen, B.; Li, W.; Yan, J.; Chen, L.; Wang, X.; Xiao, Y. Ventilator-associated pneumonia after cardiac surgery: A meta-analysis and systematic review. *J. Thorac. Cardiovasc. Surg.* **2014**, *148*, 3148–3155. [CrossRef] [PubMed]
74. Nakamura, M.; Mie, M.; Mihara, H.; Nakamura, M.; Kobatake, E. Construction of multi-functional extracellular matrix proteins that promote tube formation of endothelial cells. *Biomaterials* **2008**, *29*, 2977–2986. [CrossRef]
75. Humphries, J.D.; Byron, A.; Humphries, M.J. Integrin ligands at a glance. *J. Cell Sci.* **2006**, *119*, 3901–3903. [CrossRef]
76. Andrade, F.K.; Silva, J.P.; Carvalho, M.; Castanheira, E.M.; Soares, R.; Gama, M. Studies on the hemocompatibility of bacterial cellulose. *J. Biomed. Mater. Res. A* **2011**, *98*, 554–566. [CrossRef] [PubMed]
77. Tersteeg, C.; Roest, M.; Mak-Nienhuis, E.M.; Ligtenberg, E.; Hoefer, I.E.; de Groot, P.G.; Pasterkamp, G. A fibronectin-fibrinogen-tropoelastin coating reduces smooth muscle cell growth but improves endothelial cell function. *J. Cell Mol. Med.* **2012**, *16*, 2117–2126. [CrossRef] [PubMed]
78. Bramfeldt, H.; Vermette, P. Enhanced smooth muscle cell adhesion and proliferation on protein-modified polycaprolactone-based copolymers. *J. Biomed. Mater. Res. A* **2009**, *88*, 520–530. [CrossRef] [PubMed]
79. Shimada, T.; Nishibe, T.; Miura, H.; Hazama, K.; Kato, H.; Kudo, F.; Murashita, T.; Okuda, Y. Improved healing of small-caliber, long-fibril expanded polytetrafluoroethylene vascular grafts by covalent bonding of fibronectin. *Surg. Today* **2004**, *34*, 1025–1030. [CrossRef] [PubMed]
80. Nishibe, T.; Okuda, Y.; Kumada, T.; Tanabe, T.; Yasuda, K. Enhanced graft healing of high-porosity expanded polytetrafluoroethylene grafts by covalent bonding of fibronectin. *Surg. Today* **2000**, *30*, 426–431. [CrossRef]
81. Seeger, J.M.; Klingman, N. Improved in vivo endothelialization of prosthetic grafts by surface modification with fibronectin. *J. Vasc. Surg.* **1988**, *8*, 476–482. [CrossRef]
82. De Visscher, G.; Mesure, L.; Meuris, B.; Ivanova, A.; Flameng, W. Improved endothelialization and reduced thrombosis by coating a synthetic vascular graft with fibronectin and stem cell homing factor SDF-1α. *Acta Biomater.* **2012**, *8*, 1330–1338. [CrossRef]
83. Thomson, G.J.; Vohra, R.K.; Carr, M.H.; Walker, M.G. Adult human endothelial cell seeding using expanded polytetrafluoroethylene vascular grafts: A comparison of four substrates. *Surgery* **1991**, *109*, 20–27. [PubMed]
84. Vohra, R.; Thomson, G.J.; Carr, H.M.; Sharma, H.; Walker, M.G. The response of rapidly formed adult human endothelial-cell monolayers to shear stress of flow: A comparison of fibronectin-coated Teflon and gelatin-impregnated Dacron grafts. *Surgery* **1992**, *111*, 210–220. [PubMed]
85. Seeger, J.M.; Klingman, N. Improved endothelial cell seeding with cultured cells and fibronectin-coated grafts. *J. Surg. Res.* **1985**, *38*, 641–647. [CrossRef]
86. Osorio, M.; Ortiz, I.; Ganan, P.; Naranjo, T.; Zuluaga, R.; van Kooten, T.G.; Castro, C. Novel surface modification of three-dimensional bacterial nanocellulose with cell-derived adhesion proteins for soft tissue engineering. *Mater. Sci. Eng. C Mater. Biol. Appl.* **2019**, *100*, 697–705. [CrossRef]
87. Daum, R.; Visser, D.; Wild, C.; Kutuzova, L.; Schneider, M.; Lorenz, G.; Weiss, M.; Hinderer, S.; Stock, U.A.; Seifert, M.; et al. Fibronectin Adsorption on Electrospun Synthetic Vascular Grafts Attracts Endothelial Progenitor Cells and Promotes Endothelialization in Dynamic In Vitro Culture. *Cells* **2020**, *9*, 778. [CrossRef] [PubMed]
88. Wijelath, E.S.; Rahman, S.; Murray, J.; Patel, Y.; Savidge, G.; Sobel, M. Fibronectin promotes VEGF-induced CD34 cell differentiation into endothelial cells. *J. Vasc. Surg.* **2004**, *39*, 655–660. [CrossRef] [PubMed]
89. Bennett, J.S.; Berger, B.W.; Billings, P.C. The structure and function of platelet integrins. *J. Thromb. Haemost.* **2009**, *7* (Suppl. 1), 200–205. [CrossRef]
90. Kieffer, N.; Melchior, C.; Guinet, J.M.; Michels, S.; Gouon, V.; Bron, N. Serine 752 in the cytoplasmic domain of the beta 3 integrin subunit is not required for alpha v beta 3 postreceptor signaling events. *Cell Adhes. Commun.* **1996**, *4*, 25–39. [CrossRef]

91. Pytela, R.; Pierschbacher, M.D.; Ginsberg, M.H.; Plow, E.F.; Ruoslahti, E. Platelet membrane glycoprotein IIb/IIIa: Member of a family of Arg-Gly-Asp–specific adhesion receptors. *Science* **1986**, *231*, 1559–1562. [CrossRef] [PubMed]
92. Beumer, S.; MJ, I.J.; de Groot, P.G.; Sixma, J.J. Platelet adhesion to fibronectin in flow: Dependence on surface concentration and shear rate, role of platelet membrane glycoproteins GP IIb/IIIa and VLA-5, and inhibition by heparin. *Blood* **1994**, *84*, 3724–3733. [CrossRef]
93. Stellos, K.; Langer, H.; Daub, K.; Schoenberger, T.; Gauss, A.; Geisler, T.; Bigalke, B.; Mueller, I.; Schumm, M.; Schaefer, I.; et al. Platelet-derived stromal cell-derived factor-1 regulates adhesion and promotes differentiation of human CD34+ cells to endothelial progenitor cells. *Circulation* **2008**, *117*, 206–215. [CrossRef]
94. Stronck, J.W.; van der Lei, B.; Wildevuur, C.R. Improved healing of small-caliber polytetrafluoroethylene vascular prostheses by increased hydrophilicity and by enlarged fibril length. An experimental study in rats. *J. Thorac. Cardiovasc. Surg.* **1992**, *103*, 146–152. [CrossRef]
95. Valgimigli, M.; Bueno, H.; Byrne, R.A.; Collet, J.P.; Costa, F.; Jeppsson, A.; Juni, P.; Kastrati, A.; Kolh, P.; Mauri, L.; et al. 2017 ESC focused update on dual antiplatelet therapy in coronary artery disease developed in collaboration with EACTS: The Task Force for dual antiplatelet therapy in coronary artery disease of the European Society of Cardiology (ESC) and of the European Association for Cardio-Thoracic Surgery (EACTS). *Eur. Heart J.* **2018**, *39*, 213–260. [CrossRef]
96. Ipema, J.; Brand, A.R.; GJ, D.E.B.; JP, D.E.V.; Unl, U.C. Antiplatelet and anticoagulation therapy after revascularization for lower extremity artery disease: A national survey and literature overview. *J. Cardiovasc. Surg.* **2021**, *62*, 59–70. [CrossRef] [PubMed]
97. Braghirolli, D.I.; Helfer, V.E.; Chagastelles, P.C.; Dalberto, T.P.; Gamba, D.; Pranke, P. Electrospun scaffolds functionalized with heparin and vascular endothelial growth factor increase the proliferation of endothelial progenitor cells. *Biomed. Mater.* **2017**, *12*, 025003. [CrossRef]
98. Mazzucotelli, J.P.; Bertrand, P.; Benhaiem-Sigaux, N.; Leandri, J.; Loisance, D.Y. In vitro and in vivo evaluation of a small caliber vascular prosthesis fixed with a polyepoxy compound. *Artif. Organs* **1995**, *19*, 896–901. [CrossRef]
99. Schaeffer, U.; Tanner, B.; Strohschneider, T.; Stadtmuller, A.; Hannekum, A. Damage to arterial and venous endothelial cells in bypass grafts induced by several solutions used in bypass surgery. *Thorac. Cardiovasc. Surg.* **1997**, *45*, 168–171. [CrossRef] [PubMed]
100. Carpentier, S.; Murawsky, M.; Carpentier, A. Cytotoxicity of cardioplegic solutions: Evaluation by tissue culture. *Circulation* **1981**, *64*, II90–II95. [PubMed]
101. Zoellner, H.; Hofler, M.; Beckmann, R.; Hufnagl, P.; Vanyek, E.; Bielek, E.; Wojta, J.; Fabry, A.; Lockie, S.; Binder, B.R. Serum albumin is a specific inhibitor of apoptosis in human endothelial cells. *J. Cell Sci.* **1996**, *109*, 2571–2580. [CrossRef]
102. Zhang, W.J.; Frei, B. Albumin selectively inhibits TNF alpha-induced expression of vascular cell adhesion molecule-1 in human aortic endothelial cells. *Cardiovasc. Res.* **2002**, *55*, 820–829. [CrossRef]
103. Shen, M.; Martinson, L.; Wagner, M.S.; Castner, D.G.; Ratner, B.D.; Horbett, T.A. PEO-like plasma polymerized tetraglyme surface interactions with leukocytes and proteins: In vitro and in vivo studies. *J. Biomater. Sci. Polym. Ed.* **2002**, *13*, 367–390. [CrossRef]
104. Denizli, F.K.; Guven, O. Competitive adsorption of blood proteins on gamma-irradiated-polycarbonate films. *J. Biomater. Sci. Polym. Ed.* **2002**, *13*, 127–139. [CrossRef] [PubMed]
105. McGee, G.S.; Shuman, T.A.; Atkinson, J.B.; Weaver, F.A.; Edwards, W.H. Experimental evaluation of a new albumin-impregnated knitted dacron prosthesis. *Am. Surg.* **1987**, *53*, 695–701.
106. Fuchs, S.; Dohle, E.; Kolbe, M.; Kirkpatrick, C.J. Outgrowth endothelial cells: Sources, characteristics and potential applications in tissue engineering and regenerative medicine. *Adv. Biochem. Eng. Biotechnol.* **2010**, *123*, 201–217. [CrossRef]
107. Hur, J.; Yoon, C.H.; Kim, H.S.; Choi, J.H.; Kang, H.J.; Hwang, K.K.; Oh, B.H.; Lee, M.M.; Park, Y.B. Characterization of two types of endothelial progenitor cells and their different contributions to neovasculogenesis. *Arter. Thromb. Vasc. Biol.* **2004**, *24*, 288–293. [CrossRef] [PubMed]
108. Cheng, C.C.; Chang, S.J.; Chueh, Y.N.; Huang, T.S.; Huang, P.H.; Cheng, S.M.; Tsai, T.N.; Chen, J.W.; Wang, H.W. Distinct angiogenesis roles and surface markers of early and late endothelial progenitor cells revealed by functional group analyses. *BMC Genomics* **2013**, *14*, 182. [CrossRef] [PubMed]
109. Madhavan, K.; Elliot, W.; Tan, Y.; Monnet, E.; Tan, W. Performance of marrow stromal cell-seeded small-caliber multilayered vascular graft in a senescent sheep model. *Biomed. Mater.* **2018**, *13*, 055004. [CrossRef] [PubMed]
110. Heise, M.; Schmidmaier, G.; Husmann, I.; Heidenhain, C.; Schmidt, J.; Neuhaus, P.; Settmacher, U. PEG-hirudin/iloprost coating of small diameter ePTFE grafts effectively prevents pseudointima and intimal hyperplasia development. *Eur. J. Vasc. Endovasc. Surg.* **2006**, *32*, 418–424. [CrossRef]
111. Clowes, A.W.; Karnowsky, M.J. Suppression by heparin of smooth muscle cell proliferation in injured arteries. *Nature* **1977**, *265*, 625–626. [CrossRef] [PubMed]

MDPI
St. Alban-Anlage 66
4052 Basel
Switzerland
Tel. +41 61 683 77 34
Fax +41 61 302 89 18
www.mdpi.com

Nanomaterials Editorial Office
E-mail: nanomaterials@mdpi.com
www.mdpi.com/journal/nanomaterials

www.ingramcontent.com/pod-product-compliance
Lightning Source LLC
LaVergne TN
LVHW070648170726
843515LV00004B/352

* 9 7 8 3 0 3 6 5 5 6 1 9 2 *